American Institute of Chemical Engineers

# 4th Cape-Open US Conference

Topical Conference at the
2007 AIChE Annual Meeting

November 4-9, 2007
Salt Lake City, Utah, USA

**Printed from e-media with permission by:**

Curran Associates, Inc.
57 Morehouse Lane
Red Hook, NY 12571
www.proceedings.com

**ISBN: 978-1-60423-840-2**

**Some format issues inherent in the e-media version may also appear in this print version.**

ISBN: 978-1-60423-840-2



American Institute of Chemical Engineers
Proceedings
Three Park Avenue
New York, NY 10016-5991
Phone: 212-591-8100

www.aiche.org

American Institute of Chemical Engineers

# 4th Cape-Open US Conference

Topical Conference at the
2007 AIChE Annual Meeting

AIChE

November 4-9, 2007
Salt Lake City, Utah, USA

American Institute of Chemical Engineers

$4^{th}$ Cape-Open US Conference
2007

# TABLE OF CONTENTS

# ACHIEVEMENTS AND STRATEGY OF CO-LAN

*Dr. Werner Merk, The Dow Chemical Company, Stade, Germany*
*Ray Dickinson, Shell Global Solutions, Amsterdam, The Netherlands*
*Dr Michel Pons, CAPE-OPEN Laboratories Network, Lyon, France*

## Introduction

The development of a plug and play standard in process simulation software results from the initiative of a group of operating companies in the mid 90s. The development took the form of a series of EU-funded projects, led by IFP, starting with CAPE-OPEN (Jan 1, 1997 till June 30, 1999) and followed by Global CAPE-OPEN (July 1, 1999 till March 31, 2002) and GCO Support projects. These projects gathered operating companies, software vendors and academics. Global CAPE-OPEN was also labeled as an Intelligent Manufacturing Systems project facilitating the involvement of industrial and academic partners from Canada, US and Japan.

The CAPE-OPEN and Global CAPE-OPEN projects developed a set of interface standards that enable interoperability between the various pieces of software making up process simulation tools. The GCO Support project delivered a set of disseminating means for CAPE-OPEN standards.

Following the end of the EU projects, the CAPE-OPEN (CO) standards are now maintained and developed by the CAPE-OPEN Laboratories Network (CO-LaN), founded in 2001 by a group of twenty end user organizations, universities and software vendors. CO-LaN (www.colan.org) is funded by process industries end user companies (1).

## Goal and missions

CO-LaN is working towards its strategic goal of expanding process modeling capability through software interoperability standards.

The missions of the CO-LaN are:

1. User priorities for CO standards: work with software suppliers to clarify user priorities for interoperability, promote communication and cooperation among software suppliers and insure that the CO standards actually translate into commercially valuable interoperability.
2. Exploitation and dissemination: promote the CO standards to end-users and distribute CO information and technology internationally, so that CAPE-OPEN can be used effectively for the business operations of end-users.
3. CAPE-OPEN specifications life cycle management: organize the maintenance, evolution, and expansion of the interface specifications. This is arranged through Special Interest Groups (SIGs), each created to take care of a meaningful subset of the standards.
4. Testing and interoperability facilitation: supply compliance testers to support development of components and organize interoperability tests between software components from suppliers of CO compliant solutions.
5. Training and Migration facilitation: ensure that training modules are available and supply migration wizards to automate component wrapping. Training modules provide an effective

means of improving understanding and reducing difficulties in implementing CAPE-OPEN technology.

Most of these missions have been continuously sustained in 2006 and 2007. Meanwhile CO-LaN has continuously grown in size putting more strain on a number of missions.

## CO-LaN growth

Years 2006 and 2007 have seen a significant increase in CO-LaN membership. At its Annual General Meeting on March 7, 2006, CO-LaN counted fifty-three members, eleven more than at the same event a year before. One year later, at the AGM held on March 8, 2007, CO-LaN had sixty members. In September 2007, sixty-five members had joined CO-LaN (figure 1).

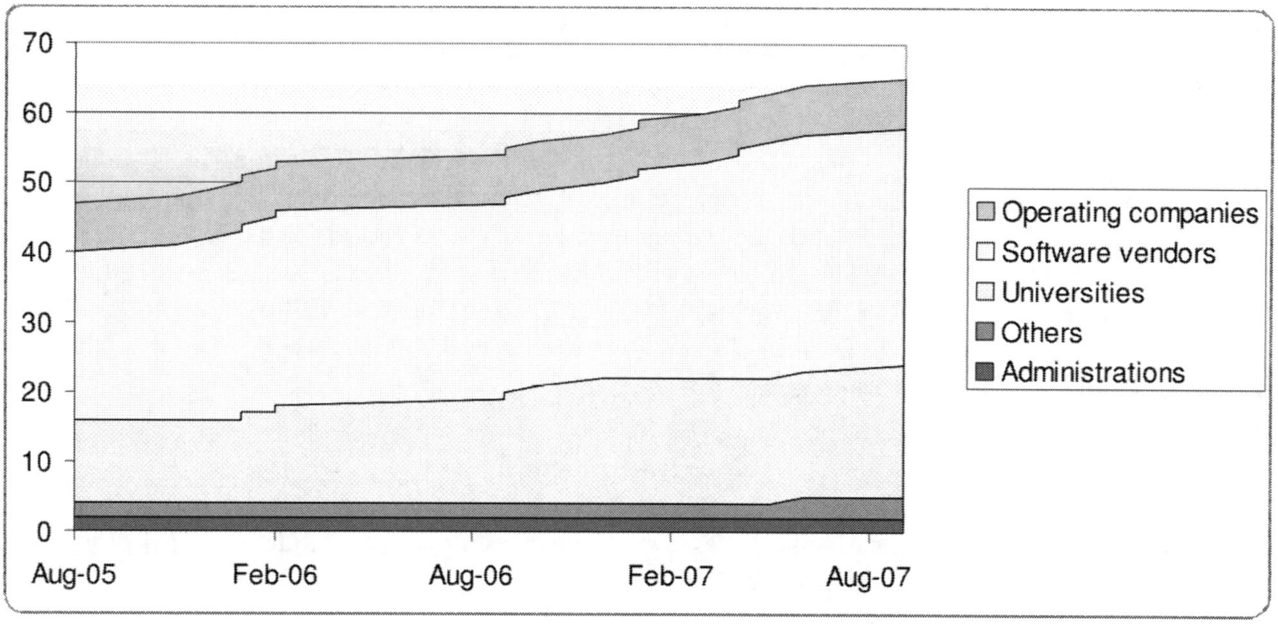


Figure 1: Evolution of CO-LaN membership

A number of US based organizations have joined CO-LaN recently (MetaHeuristics LLC, Lamar University, Iowa State University): this shows that CAPE-OPEN is less and less perceived in the US as solely a European standard. Modeling and simulation tools belong to a global market and an interface standard in this domain has to be global too. Development of the CAPE-OPEN standards was conducted by a consortium bringing in organizations from many places in the world, not only Europe. The global nature of CAPE-OPEN is being progressively accepted.

The main growth in CO-LaN membership has come from software developers from across the entire spectrum of Computer Aided Process Engineering applications. The fact that these organisations now amount to more than half of the CO-LaN membership underlines the increasing adoption of the CAPE-OPEN standards within the industry.

For the past two years CO-LaN has reinforced its structure by having a Chief Technology Officer conducting the technically based actions decided by the CO-LaN Management Board. This has resulted in much faster technical progress, as well as a higher visibility to CO-LaN, enhanced services to the membership and better support to the work of the Special Interest Groups.

The following results have been delivered to the CO-LaN membership and to the CAPE-OPEN community.

## Recent achievements

In 2006 and 2007 CO-LaN has shown progress in a number of areas.

### *User Priorities for CO standards*

CO-LaN has been in touch with many software vendors and academics to explain user priorities in terms of implementation of the CO standards. While the decision of implementing always lies on the developer side, meaning that CO-LaN is not providing any incentive for a specific implementation, it is worth noting that a large number of software tools (2) have been released with brand new CAPE-OPEN implementations or enhanced ones. New Process Modelling Environments (PMEs) are available that rely entirely on the CAPE-OPEN architecture: SolidSim from Technical University Hamburg-Harburg (3), COCO from AmsterCHEM (www.cocosimulator.org) and a process simulator from US EPA (4).

### *Exploitation and dissemination*

Achieving progress in CAPE-OPEN implementation is also a matter of bringing together software developers (experts or beginners) aiming at achieving CAPE-OPEN compliance. This is the objective of the Annual Meetings CO-LaN has been organizing both in Europe and in the US every year.

On March 9 and 10, 2006, the 3rd CAPE-OPEN European Conference brought together than 50 CAPE-OPEN practitioners, users and developers. The conference featured an Interoperability Showcase and an Interoperability Workshop plus the CO-LaN Annual General Meeting of members. The objective of the Showcase was to celebrate the progress that has been made in achieving commercial interoperability via the CAPE-OPEN standard. The Workshop provided the opportunity for users to try using CO-compliant components within commercial PMEs, for developers to try their CO-compliant Process Modelling Components (PMCs) with a range of commercial PMEs, for both users and developers to obtain expert advice on CO-compliance matters, and for training material on the mechanics of achieving CO-compliance to be shown.

On November 16 and 17, 2006, the 3rd CAPE-OPEN US Conference took place as a Topical Conference within the AIChE'06 Annual Meeting held in San Francisco, California. Comprising four technical sessions filled with more than 25 papers, the conference was highly rated by its attendees.

The 4th CAPE-OPEN European Conference took place in Heidelberg, Germany on March 8 and 9, 2007. More than sixty participants mainly from Europe but also from the US and Asia attended the sessions arranged as an Interoperability Showcase and an Interoperability Workshop. As per the survey conducted (60% response ratio), the conference contents matched closely the attendees' expectations. The topics "Meet other developers" and "Test interoperability" got the highest scores.

### *CAPE-OPEN specifications life-cycle management*

CO-LaN organizes the maintenance, evolution and expansion of the CAPE-OPEN interface specifications. In this area the major results have been:

- A Refining Reactor Special Interest Group has been launched in June 2006. Its goal is to revise and develop the necessary CAPE-OPEN interface standards for the use of Refinery Reactor models in Process Modelling Environments. The SIG key responsibilities are first to develop required extensions to existing interface specifications necessary to implement refinery reactor models as CAPE OPEN units. This may lead to the creation of new interfaces. Secondly the SIG is there to promote implementation of petroleum fractions in PMEs and PMCs. A plug prototype has been demonstrated on March 8, 2007 within a prototype version of PRO/II from SimSci-Esscor.
- A Hydrodynamic Special Interest Group has been launched in September 2007. Its creation reflects the need expressed by operating companies in the Exploration & Production market segment.
- The Thermodynamic Special Interest Group has finalized the revision of the Thermodynamic and Physical Properties interface specification. Version 1.1 of this part of the CAPE-OPEN standards has been approved for release in October 2006. It provides new capabilities and has been immediately implemented in a number of PMCs and PMEs (5).

***Testing and interoperability facilitation***

To help with PMC coding, Wizards are available. To check static compliance of PMCs, testing can be conducted either with the CAPE-OPEN Tester Suite or directly between components and environments. Once the PMC has been created, interface traffic at run time can be logged with the CAPE-OPEN Logging and Testing Tool. This set of software tools is dedicated to helping at every step of a CAPE-OPEN implementation. However, the CAPE-OPEN Tester Suite has not been updated or extended in the last two years. The maintenance process for this major piece of software is being specified in 2007 and should be implemented in 2008.

In order to facilitate interoperability testing, CO-LaN has invested in hardware dedicated to this process. Meanwhile, many software vendors have provided for free to CO-LaN licenses of their software. By remote access to this hardware, CO-LaN members are able to test their components or environments for compliance with CAPE-OPEN standards.

***Training and migration facilitation***

In its effort to reduce the developers' learning curve, CO-LaN is helping with the development of a training offer on CAPE-OPEN. Initially CO-LaN was not meant to actively participate in training sessions on CAPE-OPEN. However, since the commercial offer of such training courses was nil up to 2006, CO-LaN has taken a number of remedial actions.

In 2006 a Short Course was proposed at the AIChE'06 Annual meeting. The course was focused on CAPE-OPEN main concepts and implementation. After general background information on CAPE-OPEN, the CAPE-OPEN Common interface specifications (Identification, Parameters, etc…) were detailed. Then the Thermodynamic and Physical Properties interface specification was described with practical elements provided to help attendees evaluate how to implement the specification. The Unit Operation interface specification was subsequently discussed and illustrated with scenarios and practical examples. Finally various tools available to help developers were reviewed. The course was written and given by Alain Vacher and Stéphane Dechelotte from ProSim SA and attended by ten participants from the US, Europe and Asia. CO-LaN participated in the review of the material provided to the participants.

In 2007 CO-LaN offered on March 7 a one-day training seminar on implementing CAPE-OPEN compliant PMCs using .NET-based development tools. This seminar was geared to component developers and provided hands-on instruction. The course was a mix of presentations and on-hands coding sessions supported by the trainers. At the end of the course, participants were able to develop a mixer unit operation that was used in a COM-based PME, namely in AmsterChem's COCO. The first part of the course provided a gentle introduction to .NET. It introduced the .NET framework, highlighting aspects where it mostly differs from "conventional" programming technologies such as VB or C++. The execution of so-called managed code in virtual machines was explained and the deployment of code in assemblies containing code combined with metadata was introduced. The major part of this first section was concerned with the .NET way of language interoperability, which is a fundamental prerequisite for the interaction with environments using COM. The second section of the course introduced basics for CAPE-OPEN interoperability and was mostly concerned with the appearance of CAPE-OPEN data types in COM and .NET as well as the mapping of interface definitions. Using the development environments it was demonstrated how CAPE-OPEN interfaces can be imported into the .NET environment so that they can subsequently be used for implementation. The practical part of this course section covered the implementation of the interfaces ICapeIdentification and ICapePort for the sample mixer unit operation. The third and final section of the training course was concerned with advanced issues of implementing the CAPE-OPEN standard. After a theoretical part on the handling of collections, participants were able to develop the implementation of Port and Parameter Collections for the sample unit. Further, the automatic registration of the Unit Operation for the COM runtime was implemented. After an explanation of error handling and persistence mechanisms, participants had some time to finish the core implementation of the mixer unit. The course was attended by 20 participants. Bill Barrett from US EPA and Lars von Wedel from AixCAPE were the instructors.

The above explains what CO-LaN has done on training facilitation. Migration facilitation has been conducted through the enhancement and development of various software tools.

During the Global CAPE-OPEN project, AspenTech delivered a CAPE-OPEN Wizard for Unit Operations. Wizards are software tools building a large part of the code of a CAPE-OPEN compliant software component. The Wizard objective is to make it simpler and faster for an engineer to build a CAPE-OPEN compliant unit operation model using the Microsoft COM version of the interface standards. It is simpler because an engineer using the tool will not need to know as much about COM as an engineer trying to develop a unit operation from scratch would. And, it is faster because the tool generates a complete programming project, to which the engineer only needs to add a user interface form, a calculation routine and a validation routine.

The Unit Operation Wizard made by AspenTech and donated to CO-LaN by AspenTech is dedicated to Unit Operation written in Visual Basic. It is an add-in for Visual Basic that allows a user to create a CAPE-OPEN Unit Operation by filling in data on a sequence of forms. Once data entry has been completed the Wizard generates a Visual Basic project that builds the resulting Unit Operation. Additionally the Wizard creates an installation kit so that the Unit Operation can be easily deployed and used on other machines.

Since the work by AspenTech was completed before version 1.0 of the Unit Operation interface specification was released (5), the Wizard was compliant only with version 0.93 of the CAPE-OPEN standards. In 2006 CO-LaN contracted AspenTech to move the Wizard to compliance with version 1.0 of the CAPE-OPEN standards. The resulting piece of software has been delivered to the CO-LaN membership in October 2006.

It is worth noting that Unit Operation Wizards for programming in other languages have appeared in the meantime. IFP has donated to CO-LaN its Wizard for programming in C++ while TOTAL has donated to CO-LaN its Wizard for programming in FORTRAN 90. Both wizards are restricted to usage within the CO-LaN membership. On top of that SASOL has released as open source its own Wizard for programming in Pascal. These initiatives show the commitment of operating companies to develop PMCs internally. At the same time by making these tools available to third parties, it supports the further development of PMCs.

The CAPE-OPEN Logging and Testing tool (COLTT) is one of the tools provided by CO-LaN to help with migration and interoperability testing (6). Once its feasibility was proven by the Interoperability Special Interest Group, COLTT has been developed with a production version released on March 8, 2007. Since then COLTT has been upgraded to log PMCs compliant with version 1.1 of the Thermodynamic and Physical Properties CAPE-OPEN interface specification. COLTT is continuously maintained and enhanced by CO-LaN. Among the future enhancements planned, a change of output format is being considered, using XML to provide an easier parsing through the information logged by COLTT.

## Conclusion and perspectives

CO-LaN is now in its seventh year of activity with continuous commitment from operating companies to its funding and operation: Air Liquide, BASF, BP, The Dow Chemical Company, IFP, Shell, TOTAL are spearheading the CAPE-OPEN initiative, progressively embedding CAPE-OPEN in their work processes. CAPE-OPEN is being adopted by an ever growing group of software developers. The significant growth of CO implementations will require a strong focus of CO-LaN on life cycle management of CAPE-OPEN specifications and interoperability testing tools.

Special Interest Groups of CO-LaN are delivering documents and software for the purpose of advancing the CAPE-OPEN standards. More services and tools are continuously brought to the CO-LaN community with the objective to boost the development of more CAPE-OPEN compliant PMCs now that most PMEs have been made CAPE-OPEN compliant.

## References


1. Belaud J.P. et al. (2003), Proceedings of Foundations Of Computer-Aided Process Operations (FOCAPO), Coral Springs, Florida, USA.
2. Pons, Michel (2003), Industrial implementations of the CAPE-OPEN standard, AIDIC Conference Series, vol. 6: Selected papers of the Sixth Italian Conference on Chemical and Process engineering, pp. 253-262. S. Pierucci (ed.), Milan, Reed Business Information S.p.A., Milan.
3. Hartge, Ernst-Ulrich, M. Pogodda, Claus Reimers, Daniel Schwier, Günter Gruhn and Joachim Werther (2005), SolidSim – A new system for the flowsheet simulation of solids processes, Paper 86916, Proceedings of the 7th World Congress of Chemical Engineering, Glasgow.2.
4. Barrett, William B. and Jun Yang (2005), Development of a chemical process modelling environment based on CAPE-OPEN interface standards and the Microsoft .NET framework, Journal of Computers and Chemical Engineering, vol. 30, pp. 191-201.
5. Pons, Michel (2007), The Eight International Conference on Chemical and Process Engineering, S. Pierucci (Ed.), AIDIC, Milan.
6. Belaud, J.P. and M. Pons (2002), Computer-Aided Chemical Engineering, vol. 10: Proceedings of the 12th European Symposium on Computer Aided Process Engineering, pp 847-852, J. Grievink and J. van Schijndel (eds.), Elsevier Science B.V., Amsterdam.

# 344b Cape-Open Laboratories Refinery Reactors Sig Progress Update

**Ignasi Palou-Rivera**, Refining Technology, BP Products North America, Inc., 150 West Warrenwille Road, Mail Code 605-0W, NAPERVILLE, IL 60563-8460, Ray Dickinson, Shell Global Solutions, Amsterdam, Netherlands, David H. Jerome, INVENSYS PROCESS SYSTEMS, 26561 Rancho Parkway South, Suite 100, LAKE FOREST, CA 92630, J. Van Baten, Amsterchem, Amsterdam, Netherlands, Ensheng Zhao, Honeywell Process Solutions, Honeywell Inc., 5935 Centre Street South, Calgary, AB T2H 0C2, Canada, Michel Pons, CAPE-OPEN Laboratories Network, 32, rue Raulin, Lyon, 69007, France, Hiren Shethna, Aspen Technology, Cambridge, MA 02141, and Jean Charles De hemptinne, Institut Français du Petrole, Rueil-Malmaison, France.

**Charter:**

Revise and develop the necessary CAPE-OPEN interface standards for the use of Refinery Reactor models in Process Modelling Environments

**Key Responsibilities:**

Develop required extensions to existing interface specifications necessary to implement refinery reactor models as CAPE OPEN units. This may lead to the creation of new interfaces.

Promote implementation of petroleum fractions in PMEs and PMCs.

Deliver results back to the existing SIGs whereon this SIG can be closed.

**Deliverables for 2007:**

Engage unit and thermo SIG on modifications required for petroleum fractions.

Further refine scenarios to cover functionality required for implementation.

Initiate approval process of the petroleum fractions interface specification.

Finalize two independent prototypes for demonstration at the 2008 AGM.

# AN INTRODUCTION TO COCO

*Dr. Jasper van Baten, AmsterCHEM, Las Rozas, Spain*

## Summary

COCO is a CAPE-OPEN based steady state flowsheet simulation environment. COCO consists out of 4 CAPE-OPEN based software components: a flowsheeting environment (COFE), a thermodynamic system (TEA), a set of unit operations (COUSCOUS) and a reaction package manager (CORN). COCO is created with CAPE-OPEN interoperability testing and development in mind, but features a fully functional flowsheet simulation environment. TEA comes with an extensive thermodynamic model library, available in CAPE-OPEN thermodynamic standard versions 1.0 and 1.1. The thermodynamic property calculations and compound database are based on over 10 year experience of the distillation column simulation package ChemSep (http://www.chemsep.com/).

Whereas other simulation environments offer access to simulation components via CAPE-OPEN interfaces, COCO is modeled around CAPE-OPEN interfaces. The flowsheeting environment does not have integrated unit operation models or thermodynamic calculations: all such calculations are done only via CAPE-OPEN interfaces. This setup allows for using each of the COCO components in combination with any other equivalent CAPE-OPEN compliant components.

COCO can be downloaded for free at http://www.cocosimulator.org/. Within the CAPE-OPEN community, COCO is extensively used for interoperability testing. As a result of its free availability and extensive usage, COCO frequently receives feedback on CAPE-OPEN interfacing and interoperability issues. For the COCO development team this provides a means to remain up-to-date with the latest CAPE-OPEN developments; the COCO developers actively take part in CAPE-OPEN Laboratories Network activities and especially in its various Special Interest Groups (SIGs).

## Introduction

A typical simulation environment supporting CAPE-OPEN[1,2] Process Modeling Clients, or PMCs, is sketched in Figure 1:

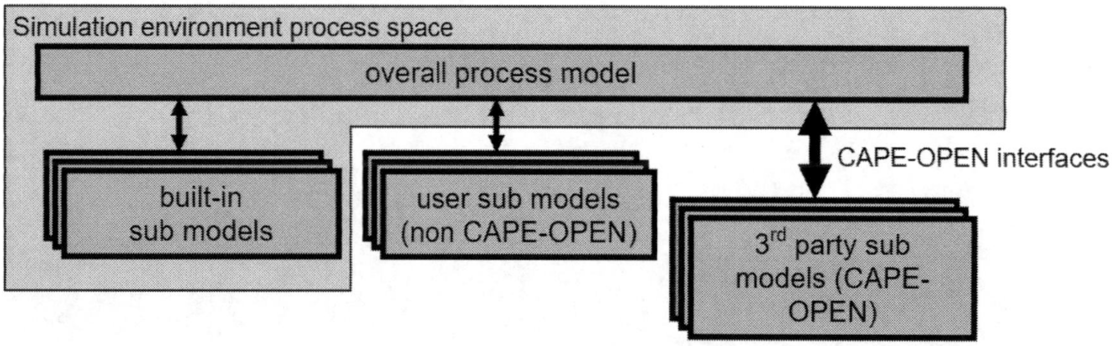


Figure 1: typical setup for process engineering modeling tools rting CAPE-OPEN models.

COCO is a steady state simulation environment, for CAPE-OPEN interoperability testing and development as well as for simulation purposes. The setup of COCO is much like the typical setup in Figure 1, except for that COCO's flowsheeting environment COFE does not require any built-in sub models or thermodynamics or proprietary interfaces; COCO is entirely modeled around CAPE-OPEN. Therefore, you can run COFE with COCO's thermodynamics library TEA, or with any third-party CAPE-OPEN compliant thermodynamic components; you can run COFE with COCO's unit operation models COUSCOUS, or with any other third-party CAPE-OPEN compliant unit operation models. Similarly, you can use COCO's thermodynamics TEA or COCO's unit operation models COUSCOUS in COFE, but also in any other third-party CAPE-OPEN compliant simulation environment.

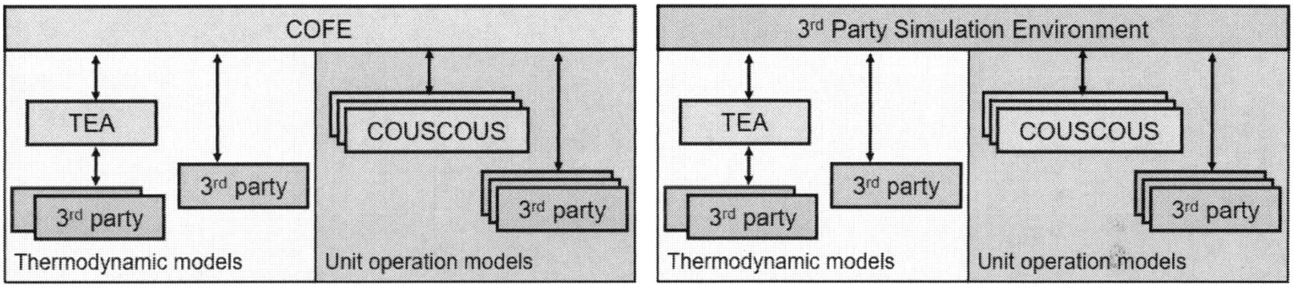

Figure 2: Schematic representation of simulation environments with thermodynamic models and unit operation models. All of COCO's components are exchangeable. Support for external models as indicated for third-party simulation environments differs from product to product.

At this moment, all COCO's CAPE-OPEN interfaces are COM interfaces, and therefore run on Microsoft Windows® platforms[3]. COCO received the CO-LaN CAPE-OPEN 2006 Award.

## COFE: CAPE-OPEN Flowsheeting Environment

COFE is a steady state flowsheeting environment. It has an intuitive graphical front-end for flowsheet building. Steady state flowsheets are solved by partitioning, followed by cutting and tearing for partitions containing recycles. Recycles are detected by structurally traversing the partition's streams, considering groups of unit operations as a single entity. After removing self-recycles from the equation, recycle tearing aims for an optimal tearing set by selecting tear streams on criteria like cutting recycles as few times as possible, cutting as few variables as possible, and having as few cut streams as possible. Recycles are solved using a hybrid approach, switching between Wegstein substitution and full Newton iterations.

COFE supports CAPE-OPEN thermodynamic property packages (versions 1.0 and 1.1), CAPE-OPEN unit operations and CAPE-OPEN reaction packages. Configuration data of each CAPE-OPEN PMC is easily accessed, either through CAPE-OPEN parameter interfaces or using the model's native editing methods.

COFE features property graphing, a built-in property calculator, multiple material types allowing for sub-sets of compounds and easy switching of thermodynamic models, and standard Windows features like graphical printing and print preview, OLE support, copy / paste, as well as an extensive on-line help. An Automation interface allows accessing COFE flowsheets in a programmatic manner; a direct application hereof is the ability to incorporate flowsheets in a Microsoft Excel workbook, and perform thermodynamic calculations as well as access stream and unit operation data.

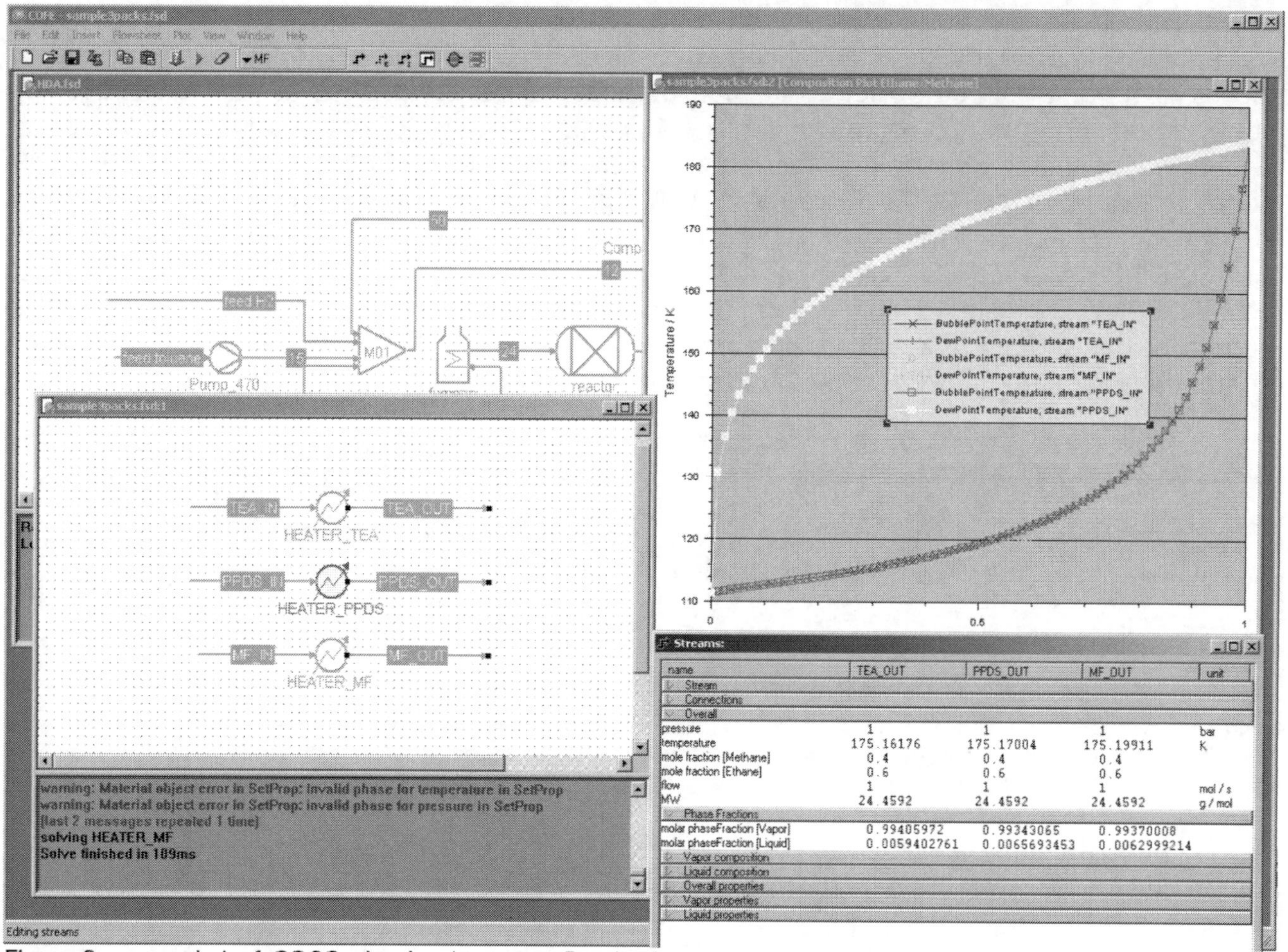


Figure 3: screenshot of COCO showing two open flowsheet documents. The document on top shows comparison of thermodynamic calculations of three different vendors. The stream view on the bottom right lists the results of the stream exiting the heaters (with equal configuration and equally defined inlet streams). The plot on the top right shows the phase envelopes for the three thermo systems at inlet stream conditions.

COFE is built not only for simulation purposes, but also with CAPE-OPEN testing and development in mind. Therefore, COFE does extensive error checking and logging. Errors and warnings are collected separately for each unit operation at the last iteration of their solution. CAPE-OPEN interface errors clearly show in red text in the log window that sits at the bottom of every flowsheet document. Messages sent to COFE by external models show in blue in the log window. Information messages show in black. Critical CAPE-OPEN errors are shown in a message box.

Figure 3 shows a screen shot of COFE. Figure 4 shows an example of COFE's Automation; running COFE in Microsoft Excel.

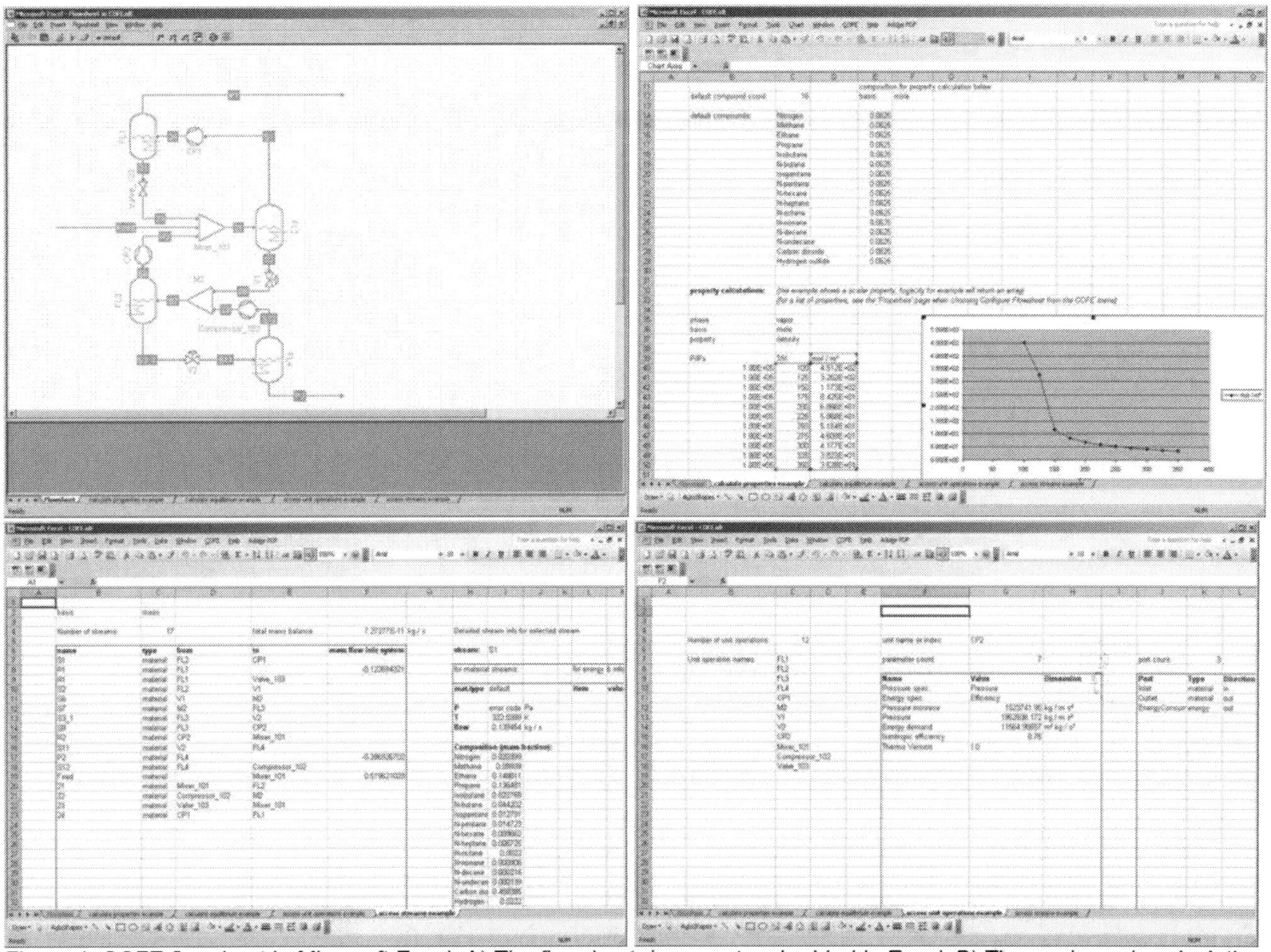

Figure 4: COFE flowsheet in Microsoft Excel. A) The flowsheet document embedded in Excel. B) Thermodynamic calculations utilizing the flowsheet's thermodynamic setup C) accessing stream data from the flowsheet D) accessing unit operation data from the flowsheet.

## TEA: Thermodynamics for Engineering Applications

TEA is COCO's thermodynamic calculation engine. It implements a CAPE-OPEN version 1.0 Thermo System as well as a CAPE-OPEN version 1.1 Property Package Manager. With an extensive library of property calculation routines, TEA exports more than 25 different properties with their pressure-, temperature- and composition derivatives. Each property calculation method can be selected individually, allowing the user to pick from more than 100 property calculation routines. TEA comes with a pure compound database that can be extended with user data or by using commercially available compound databases (DIPPR). TEA includes binary interaction coefficients for various models, which the user can override. Also all compound properties and coefficients to correlations can be overridden by the user. Most of the calculation routines that are built into TEA are based on the implementation of ChemSep (http://www.chemsep.com/), a distillation column simulation program. In doing so TEA shared in more than 10 years of experience of the ChemSep development team. Pseudo properties are implemented for infeasible solution regions of cubic equations of state[4], and TEA comes with a built-in vapor-liquid equilibrium calculator using inside-out flash calculations[5,6], with as backup a full Newton solution to flash calculations.

TEA supports plugging in external property Calculation Routines (CAPE-OPEN version 1.0) as well as external Property Calculation Routines and Equilibrium Calculation Routines

(CAPE-OPEN version 1.1). This allows overriding property calculations that are done by TEA by any other third-party calculation server, as well as overriding the equilibrium model. As a result, you can use TEA for your thermodynamic calculations, but you can also use TEA as a means to combine multiple third-party property packages or calculation routines to define a custom set of thermodynamic property calculations.

**COUSCOUS: CAPE-OPEN Unit operations (Simple)**

COUSCOUS is a library of unit operations. COUSCOUS aims to allow for the most common process engineering calculations. Most of the COUSCOUS models are simple and straight-forward, including unit operations such as compressors, pumps, turbines, expanders, simple reactor models, flash unit operations, heaters, coolers, heat-exchangers, mixers and splitters. COUSCOUS includes a no-operation unit operation, which is intended for testing CAPE-OPEN operation only. More-over, it contains a unit operation specifically to test property package implementations. This unit operation will systematically calculate all properties and compare obtained property derivatives with derivatives that are calculated by perturbation. Unit operations like this aid in CAPE-OPEN development and are not intended to contribute to process simulations. All COUSCOUS' unit operations can deal with CAPE-OPEN thermodynamic versions 1.0 and 1.1.

COUSCOUS does not include a distillation column. COCO however ships with a LITE version of ChemSep; this allows for calculation equilibrium distillation columns with a limited number of compounds and stages. A full version of ChemSep, featuring reactive distillation and non-equilibrium calculations, can be purchased separately.

**CORN: CAPE-OPEN Reaction Numerics**

CORN is COCO's reaction packages; it features kinetic and equilibrium reactions. For reaction kinetics, equilibrium constants and heats of reactions, formulas can be entered depend on pressure, temperature, concentration, mole fraction or activity. CORN deals with CAPE-OPEN thermodynamic versions 1.0 and 1.1.

Currently, there is not a wide support for reaction packages. The main application for CORN at the moment therefore is to run COUSCOUS reactors in COCO. CORN can aid in development of CAPE-OPEN simulation environments that aim to implement support for reaction packages.

**COCO: CAPE-OPEN to CAPE-OPEN environment**

COCO is the collection of COFE, TEA, COUSCOUS and CORN. In addition to that, it contains utilities like CORK (CAPE-OPEN Resource Kit), a utility to inspect and manipulate CAPE-OPEN object registration. The COCO distribution also includes third-party contributions. Currently, it includes the LITE version of the ChemSep distillation column simulation program (http://www.chemsep.com/) and COSMO*Logic*'s activity calculation routine COSMO*Therm* (http://www.cosmologic.de). Developers and software vendors are encouraged to include freely available, functional, CAPE-OPEN compliant software components in the COCO distribution.

**Interoperability**

COCO comes with a number of test-problems (available from the web site), such as the Hydro De-Alkylation plant[7], and the Cavett test-problem for flowsheet convergence[8]. Each of these test problems can be used to insert any other thermodynamic model or unit operation for interoperability testing.

COCO and its components have been submitted to extensive cross-testing against third-party products. Often these tests are successful; sometimes the tests expose items that are amendable for improvement. As a testing platform, COCO is popular within the CAPE-OPEN community. As a result, COCO often gets feedback on things that do or do not function properly in inter-model communication. Sometimes this leads to discussions that involve multiple software vendors and the CAPE-OPEN special interest groups (SIGs) on how the standard or its documentation can be improved upon. COCO has recently launched a CAPE-OPEN interoperability testing program. For more details, please see http://www.cocosimulator.org/.

Shown in Figure 3 is a test example in which three different thermodynamic systems; COFE allows for loading multiple thermodynamics sets at the same time. You can create multiple types of material definitions, and assign a thermodynamic property package to each material definition. Cross-testing can be done by setting up the problem twice (as in Figure 3) or by re-assigning the thermodynamic set to the material type and rerunning the problem. An example of TEA thermodynamics running in AspenPlus 2004.1 is shown in Figure 5; in this example TEA imports activity coefficients from COSMO*Therm*.

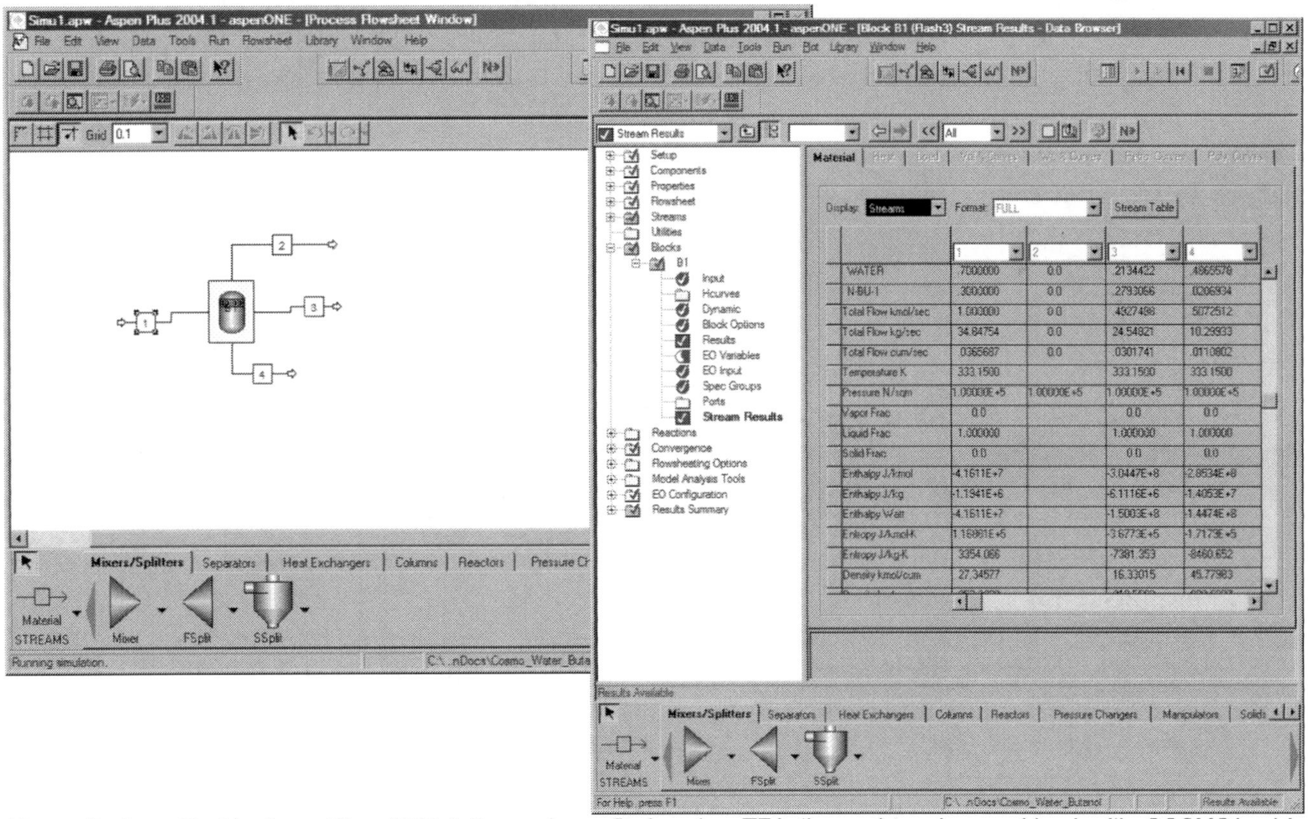


Figure 5: AspenTech's AspenPlus 2004.1 three phase flash using TEA thermodynamics combined with COSMO*Logic*'s activity calculation routine COSMO*Therm*, as presented by Frank Eckert at the annual CAPE-OPEN conference in Cannes (France), March 2006.

In allowing each of the software components to be used and tested in any desired environment, COCO makes it possible for developers without access to proprietary simulation environments to reliably develop new process models and thermodynamic models. Because of the activities of the COCO developers in the CAPE-OPEN special interest groups and frequent discussions with software developers, COCO contributes to enhancements and additions to the CAPE-OPEN standard specifications and its documentation. By means of its recent testing program, COCO contributes to the steadily improving quality of existing CAPE-OPEN implementations currently available, and promotes new implementations to be released.

### Availability

COCO is available for download free of charge from http://www.cocosimulator.org/.

## Conclusions

COCO is a collection of CAPE-OPEN software components for steady state process simulation and CAPE-OPEN interoperability testing. It is the award-winning simulation environment that is entirely based on CAPE-OPEN, allowing developers and industry to fully test and exploit the power and versatility of CAPE-OPEN technology.

COCO is available free of charge. It contains all requirements for basic steady state flowsheet simulations, and is therefore suitable for academic use and educational courses on thermodynamics, reactions, separations and process design and synthesis.

CAPE-OPEN builds bridges between commercial industrial software and academia for development of dedicated process software in reactor engineering and separation technologies. COCO contributes because of its availability and because is it open to contributions from third parties.

## References

1. Braunschweig, Bertrand L., Pantelides, Constantinos C., Britt, Herbert I. and Sama, Sergi (2000). Process modelling: the promise of open software architectures, *Chemical Engineering Progress*, September, pp 65-76.
2. Barrett, William M. and Yang, Jun (2005), "Development of a chemical process modeling environment based on CAPE-OPEN interface standards and the Microsoft .NET framework". *Computers and Chemical Engineering* 30, pp. 191-201
3. Microsoft (1995). The component object model specification. Redmond: Microsoft Corporation.
4. Mathias, P.M., Boston, J.F., Watanasiri, S. (1984), "Effective Utilization of Equations of State for Thermodynamic Processes in Process Simulation", AIChE Journal, vol 30 no 2, pp 182-186
5. Boston, J.F. and Britt, H. I (1978), "A radically different formulation and solution of the single-stage flash problem", *Computers and Chemical engineering*, 2, pp 109-122
6. Parekh, Vipul S. and Mathias, Paul M. (1998), "Efficient flash calculations for chemical process design - extension of the Boston-Britt Inside-out flash algorithm to extreme conditions and new flash types", *Computers and Chemical engineering*, 22, pp 1371-1380
7. Douglas, J.M. (1985), "A hierarchical decision procedure for process synthesis", *AIChE Journal* 31 pp. 353-362.

8. Cavett, R.H. (1963), "Application of Numerical Methods to the Convergence of Simulated Processes Involving Recycle Loops", *American Petroleum Institute*, 43, p 57

## Acknowledgements

The COCO team gratefully acknowledges the CO-LaN organization, and the software vendors that have supplied the COCO development team or AmsterCHEM with software licenses allowing for interoperability testing; specifically ChemSep (http://www.chemsep.com/), COSMOLogic (http://www.cosmologic.de/), AspenTech (http://www.aspentech.com/), Infochem (http://www.infochemuk.com/), SIMSCI-ESSCOR (http://www.simsci-esscor.com/), TUV-NEL (http://www.tuvnel.com/), PSE (http://www.psenterprise.com/) Heat Transfer Research, Inc (http://www.htri.net/).

# 344d Application of Cape-Open Standards for the Interoperability between Computer-Aided Modelling Tools Mot and Process Simulators Prosimplus

**Ricardo Morales-Rodríguez**[1], Rafiqul Gani[1], Alain Vacher[2], Pierre Castelain[2], and Stéphane Dechelotte[2]. (1) Chemical Engineering, Technical University of Denmark, 2800 Kgs. Lyngby, Denmark, (2) ProSim, Stratege Batiment A, BP 2738, Labege Cedex, 31312, France

It is well known that computer-aided tools (software) play a vital role in the modelling/design/analysis/control of chemical products and processes that manufacture them. As these tools may come from different sources and disciplines, an important issue is how they can be used efficiently for the design/analysis of specific process-product? One alternative is to employ the CAPE-OPEN interface standards for the integration of diverse computational tools.

The objective of this presentation is to highlight the application of computer-aided modelling tools for the generation and use of modelling objects in CAPE-OPEN compliant process simulators. This presentation will highlight the use of external modelling objects for unit operation models as well as thermodynamic property models. We will illustrate the following systems:

• A process simulator with a generic CAPE-OPEN socket (e.g. ProSimPlus).

• Simulis Thermodynamics to test the thermodynamic calculations.

• A CAPE-OPEN unit/thermo plug/wrapper (link) available to receive an external model object from an external source.

• The model object representing any model (unit operation or thermo) generated through a computer-aided modelling tool for model translation, analysis and solution (ICAS-MoT).

The advantage of using ICAS-MoT is that it is able to transfer the model equations representing any new unit operation/thermo into a COM-object that is able to be used in various simulation engines (Icassim and Dynsim available within ICAS software), external software (i.e. Excel) or external simulators (ProSimPlus, etc) with a well defined interface without the user having to write any programming code. On the other hand, ProSim has Simulis Thermodynamics and ProSimPlus; where, Simulis Thermodynamics is a thermophysical properties calculation server available as an MS-Excel add-in, a MATLAB toolbox or as a module pluggable in any other software that requires thermophysical properties. Simulis Thermodynamics is designed in particular for calculation of any type of transport properties (specific heat, viscosity,...), thermodynamic properties (enthalpy, compressibility factor,...) or phase equilibria (VLE, LLE, VLLE, salt formation,...) on multi components mixtures. ProSimPlus is a process engineering software that performs rigorous ss and energy balance calculations for a wide range of industrial steady-state processes. It is used in design as well as in operation of existing plants for process optimization, units troubleshooting or debottlenecking, plants revamping or performing front-end engineering analysis.

The presentation will illustrate the application of the CAPE-OPEN standards for a model generated and tested in ICAS-MoT. First the equivalent COM-object is created in ICAS-MoT. This COM-object is then wrapped with a unit-wrapper and plugged into a general CAPE-OPEN unit operation socket that is available for any simulator to be used either from MoT or ProSimPlus. It will also highlight the synergy between ICAS-MoT and Simulis Thermodynamics through the use of a plug/wrapper socket for the calculation of thermodynamic properties based on externally supplied property models. The main objective with the application examples is to highlight the generic nature of the CAPE-OPEN plug and socket that allow smooth and easy integration.

# 344e Cape-Open Implementation in Chemcad: Thermo Property Package Sockets, Integrated Pmc Generator, and Unit Operation Socket

**Yimin Peng** and **Neil H. Gray**. Chemstations Inc., 2901 Wilcrest Drive, Houston, TX 77042

This paper presents the implementation of CAPE-OPEN interfaces in the powerful process simulation suite, CHEMCAD. Examples of interoperability with CAPE-OPEN PMC's will be given.

The implementation of CAPE-OPEN Thermodynamic and physical property interface enables CHEMCAD users to access third party CAPE-OPEN thermo property packages within CHEMCAD with only a few mouse clicks. The implementation includes the equilibrium flash routine, K-values, log-fugacity coefficient, fugacity-coefficient and many other basic thermo or physical property routines. A log tool is included for tracing all calls to a package.

The integrated PMC generator in CHEMCAD enables users to generate CAPE-OPEN compliant Thermo property package easily from any of the large number of thermodynamic models and chemical components built into the program. Best of all, an entire process simulation environment is available, which includes tools for generating, plotting, and regressing equilibrium data, enabling fast and easy determination of optimal thermodynamic model(s) for a chosen system.

Using CAPE-OPEN unit operations in CHEMCAD is also designed to be intuitive. The user's interaction with these unit operations follows the familiar interface of native CHEMCAD unit operations. The selection, configuration, and use of CAPE-OPEN unit operations conforms to the same interface styles that CHEMCAD uses. Output from CAPE-OPEN operations fits into existing flow sheet reporting tools. CAPE-OPEN unit operations integrate into CHEMCAD's sequential, simultaneous, and dynamic calculation modes.

# Class-Based Cape Open Interface Revisited In .Net Framework

Maurizio Fermeglia and Marco Parenzan
DICAMP-MOSE, University of Trieste
Piazzale Europa 1, 34127, Trieste, ITALY
mauf@mose.units.it; marco.parenzan@libero.it

## Abstract

CAPE-OPEN standard is defined as a series of interfaces described in a formal documentation set. They are based on universally recognized software technologies developed in the nineties, such as COM and CORBA. Version 1.0 of CAPE-OPEN specifications is implemented in COM (Component Object Model by Microsoft). In the last years, Microsoft has designed and developed a new specification for interoperability among different platform for implementing 'cross platform software', namely .NET.

The .NET framework is definitely a step forward into interoperable and easy-to-develop applications. Since .NET is the evolution of COM technology, it allows the execution of COM codes and the integration of COM code and native .NET code. COM interoperability is the fundamental runtime functionality that allows traditional COM applications using components developed with .NET languages.

In this paper we present the design and the implementation of a software layer developed in .NET framework with C#. This software layer talks with the standard CAPE-OPEN COM interfaces, but allow a much more efficient development of codes for the process engineers using standard .NET environment. All this, without modifying the CAPE-OPEN standard interfaces included into most process simulation software available in the market today.

With the proposed framework it is possible to develop cape open unit operation and thermodynamic models in a much simpler and self-evident way making it easier for non specialist to write the CO necessary code. Examples are reported of unit operations and thermodynamic property modules developed in the new framework.

## Introduction

The user who has to design, for example, an unit operation in a process simulator needs a clear, simple and self-descriptive schema, in which it is possible to implement the equations without wasting too much time understanding the code-structure and the software infrastructure and architecture. For this reason, the user, namely the process engineer that has to develop a new unit operation, should be able to ignore the effective implementation of CAPE-OPEN (1) interfaces, and to focus only on (few) methods that the programmer has decided to show him. These methods will have to be strongly standardized so that the user can immediately understand their use and their functionalities, in terms of process engineering, directly by looking at the name of the methods and the returned value.

To achieve this goal, it is necessary to "hide" the complexity of CAPE-OPEN interfaces to the process engineer, so that she can concentrate on the equations to be developed. To do this, it is necessary to separate the implementation of interfaces, which must be unique for all the different projects, from the specific part of code of the unit operation to be developed. By doing so, the user will have at her disposal only two methods of a given class, which will serve as "constructor" of the unit and as "calculate method". This corresponds to a simple and rather clear schema: one method to initialize and another to make it run. Lastly, if she needs a more complex and "powerful" structure, she could anyway benefit by the object-oriented model.

The .NET framework (2) allows the developer to use typical and powerful issues of object orientation such as class inheritance and strong typing.

Class inheritance allows the development of new classes inheriting code from parent classes. Many times implementation of interface methods is a repetitive work due to infrastructure needs rather than to a specific need. Class inheritance allows the building of a class framework minimizing the work needed to code a fully functional class.
Strong Typing is the ability for a compiler to check all types' compatibility at compile time ("early binding"). This allows avoiding runtime problems such as "Invalid Cast Exception" due to an extensive usage of Variant (3) or void* (C/C++).
Besides this, the possibility of making extensive use of Intellisense in development tools such as Visual Studio .NET makes coding a faster and more productive experience.

This paper presents the architecture, design and the implementation of a software layer developed in .NET framework with C# with the objective of simplifying the development of process simulation modules in the Cape Open framework. The software developed is based on Cape Open, namely it fully respects the Cape Open specifications at the interface with process simulators, but exposes simpler and self explaining interfaces to the user. One objective of the new software tool is the separation of the infrastructure with the equations describing material and energy balances, thus allowing the user to concentrate on the latter and, only occasionally and for specific purposes, to modify the infrastructure. The software layer described in this work talks with the standard CAPE-OPEN COM interfaces, but allow a much more efficient development of codes for the process engineers using standard .NET environment. All this, without modifying the CAPE-OPEN standard interfaces included into most process simulators available in the market today. With the proposed framework it is possible to develop cape open unit operation and thermodynamic models in a much simpler and self-evident way making it easier for non specialist to write the CO necessary code. Examples are reported of unit operations and thermodynamic property modules developed in the new framework.
This paper completes and extends the prototype developed and presented previously (4, 5) and adds examples and applications. The idea to develop a software layer to simplify the development of Cape Open modules came up when, some years ago, we looked at the problem from the user point of view. There was a need: to implement a CAPE-OPEN unit operation and the most obvious choice was to use the wizard. Very soon, by looking at the code generated by the wizard, we realized that this was not a good solution for a process engineer that has to concentrate on writing the equations of a given CO unit: too much plumbing in VB6-generated code; too many variants; too much COM for a process engineer.

Fortunately, in our group we have a long tradition of software development and we are .NET developers since 2001 (first beta of .NET), so we decided that .NET could solve most of the problems encountered in coding directly in COM environment using VB6 and the wizard. In fact .NET overcomes problems of VB6 to handle full object orientation and strong typing, very useful to manage complex projects. Coding in C++ with MFC or ATL was not the solution: too complex and plumbing-dependent for projects with limited time-frame, such as a CAPE-OPEN module. In summary, .NET with native COM interoperability has been selected as the pillar for the new development: this choice will allow us to make an extensive use of Object Oriented Programming paradigm (6) to develop our software. It allow to encapsulate infrastructure code into an entity called "object" that has a well defined structure called "class". More specifically "objects" are "instances" of a class.
We decided to start a project for the migration of the VB6 code to .NET and then subsequently re elaborate the API with the objective to simplify the development for process engineers.

## The Object Oriented Programming Paradigm

Perhaps the biggest issues that becomes evident when looking at the source code generated by the CAPE-OPEN Code generated by the VB6 wizard is the great amount of infrastructure/repetitive code. The fact that is automatically generated does not imply that it is not a matter of the programmer. In fact, the main problems related with the code are:

- Difficulties to understand what the code is doing;
- Difficulties to find the real point where to add code to the infrastructure;
- Difficulties to manage the project (many files, most of which remains unmodified);

We analyzed the existing VB generated code in terms of meaning and reusability:

- What is this code about?
- Which code depends on specific wizard generation?
- Which code is independent from wizard generation?

This analysis allowed us to evaluate (roughly) that only 20% was specific code, the rest being independent from the specific unit operation development.
In terms of CAPE OPEN, the concept of a Unit can be modeled into a "class". Each Unit class contains a collection of ports and a collection of properties. Each Unit has the same kind of collection of ports or classes. When we need to use a Unit, an instance of a Unit class will be created and an object of type "Unit" is used. There is an issue on Calculate code, that in OOP terms we call a member function: the code is specific to each Unit. This implies that there cannot be a single Unit class, but many Unit classes are needed.

*Source Code Generation and Inheritance*
VB6 wizard approach for unit code is a Source Code Generation approach (7). Inheritance is the OOP principle by which a class can be defined as "derived from" another class: all the code defined in the "super" class , from which we inherits, is already available to the newly defined derived class. As a simple effect, in the source code file of the derived class the code of the base class is not present: there is only a reference to the super class. This permits the generate code whose readability is much enhanced.

*Polymorphism*
Each unit class must have a Calculate function. The base unit class that contains all the code that can be now inherited into all the newly derived classes, must contain a Calculate function to be considered a unit. But what specific code is contained into the Calculate function of the base, generic, unit class? The most obvious answer is "nothing": the calculate function in the base unit class is empty. So: how to solve this? There is a special condition, called "virtual function" (8) by which a function, already defined in the base class, can be redefined (overridden) in the derived class.
A simple example can explain this issue better.
The code below is written in Microsoft Visual Basic .NET (9), an evolution of the original Visual Basic (3), re-implemented into the .NET Framework (2).

```
'
'   The base class in which define all the code
'   common among all Unit classes
'   For example the collections of Ports and Parameters
'
Public Class UnitBase

    Protected Ports As Collection
    Protected Parameters As Collection

    Public Overridable Sub Calculate()

        '
        '   do nothing
        '

    End Sub

End Class

'
'   The specific unit that inherits Ports and Parameters collection
'   Note that MyUnit inherits from UnitBase
'
Public Class MyUnit
    Inherits UnitBase

    Public Overrides Sub Calculate()
        MyBase.Calculate()

        '
        '   do something
        '

        Ports("OUT").Value = Ports("IN").Value

    End Sub

End Class
```

*COM and OOP, VB6 and VB.NET*

Now that a Unit Operation is defined in this new way, how is the relationship with CAPE-OPEN established? We have to remember that our approach does not change the CAPE-OPEN specifications, the specifications are only "hidden". At this point, one issue should be kept in mind, namely that COM, the technology behind "classic" CAPE-OPEN development, is entirely based on polymorphism. In fact, in simple mode, a COM interface (and so all the CAPE-OPEN interfaces) can be described as "a class which defines functions that have to be overridden in the implementing class". The significance of this is that, for example in VB6 environment, one has to create a ".cls" class file, whose first line is typically

```
Implements ICapeUnit
```

Then, the VB6 editor defines some function code for each function defined in the interface. That's it. With interfaces, the programmer is constrained to define all the functions, and, in VB6, the programmer does not have any method to reuse some code. In other words, VB6 misses inheritance, more specifically it misses class inheritance, because interface implementation is called also "interface inheritance".

*COM, PMC and PME*

The definition of an interface, or a base class, defines the relationship between Process Module Component (PMC) and Process Module Environment (PME). The Unit operation to be developed is the PMC, a Process Simulator is a PME.
A PME needs an interface, or a base class, because it must have to know some functionalities, for example the Calculate function. In terms of our example above, a PME can be represented in the following way:

```
Public Module PME

    Sub Main()

        Dim unit As UnitBase
        unit = New MyUnit

        unit.Calculate()

    End Sub

End Module
```

So, a true CAPE-OPEN PMC must implement ICapeUnit interface and all other interfaces required with the same mechanism. Our UnitBase must be connected to the CAPE-OPEN "classic" interfaces and, since our framework is implemented in Microsoft .NET (2), we can access the entire functionality of "COM interoperability" (10) that is described in (11, 12). Our UnitBase must implement the COM interface ICapeUnit as follows:

```
Public Class UnitBase
    Implements CAPEOPEN093.ICapeUnit

    Protected Ports As Collection
    Protected Parameters As Collection

    Public Overridable Sub Calculate() Implements CAPEOPEN093.ICapeUnit.Calculate

        '
        '   do nothing
        '

    End Sub

#Region " Other ICapeUnit functions "

#End Region

End Class
```

With the standard support from the .NET Framework, any class can be registered as a COM class, included a derived class from out UnitBase class.

Figure 1 shows the general architecture of the software solution described. The PME is on the right of the schema, while the code to be written by process engineers is on the left of the figure. The central part in figure 1 depicts clearly the role of the framework developed in this paper.

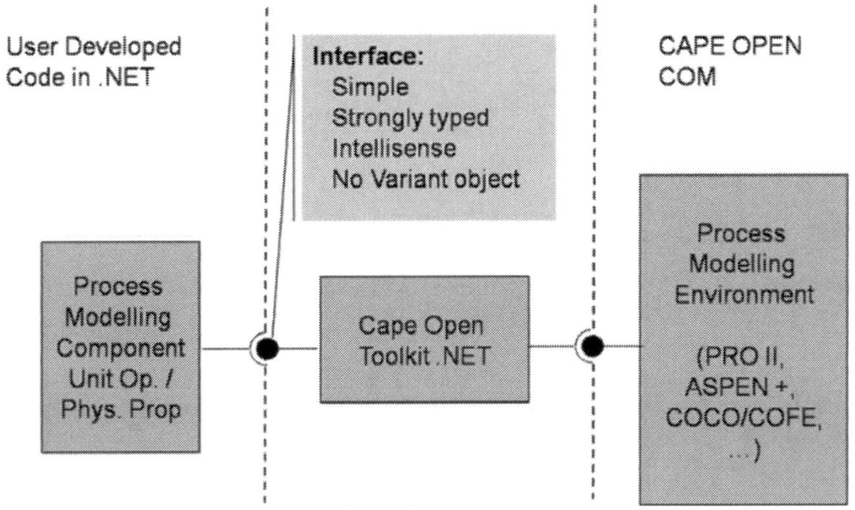


Figure 1: the software architecture of the solution proposed.

*The class based framework*

The most important interface in the CAPE-OPEN standard is the ICapeIdentification interface. It is the basis for each structure and it requires a name and a description (Units, Ports, and Parameters). Other common elements are some types of error which can occur in each class: ECapeUser, ECapeRoot. Each class therefore must contain an object which allows the error-handling.

Each class will then be derived from the class which implements ECapeUser and ECapeRoot, and the classes describing Units, Ports or Parameters will derive from a class which contains also ICapeIdentification implementation.

**Port**. The concept of "Port" doesn't describe an actual element. In other words it represents a vast set of very heterogeneous elements, even if with similar characteristics. Therefore, it is not possible to work on a Port if the type it represents is not known. It is then an abstract element which provides the structural basis for each of its specializations (Material Port, Information Port, and Energy Port).

**Parameters**. The analysis is similar to the previous case: the abstraction of Parameter exists, but the actual element is always a specialization (Integer Parameter, Real Parameter, etc.). This distinction is implicit in the ICapeParameterSpec interface, to which the UnitParameter class refers. Therefore, the management of domain and parameter value is already standardized, thanks to this device, just like in the original VB6 project. To hide the CAPE implementation to the user it is anyway convenient to manage the Parameters as different entities.

**Carried material**. If the Unit operation has Material Ports, they have material fluxes that will be accessed by ICapeMaterialObject interface. This interface exposes, for the setting and the return of property, methods which can be distinguished on the basis of complexity of syntax and of unnaturalness of the sort of returned value. It would thus be convenient to encapsulate these methods in other functions and properties, thus hiding the effective implementation and converting the type of returned value.

**Unit**. The developer of custom components needs to access only the constructor of the class, which represents the Unit, and the Calculate method. It is therefore possible to show the programmer only these two methods. The effective implementation of all the other classes could, and should, be hidden. Consequently, the Unit becomes an abstract concept and the Custom Unit then becomes the real entity k on. A class is defined, which represents conceptually

the Unit and implements CAPE-OPEN interfaces. Another class is defined which allows declaring Ports, Parameters and the Calculate method of the specific Unit.
The resulting class-based framework is described in figure 2. All classes are contained in a .NET assembly that needs to be referenced in each unit project to be developed.

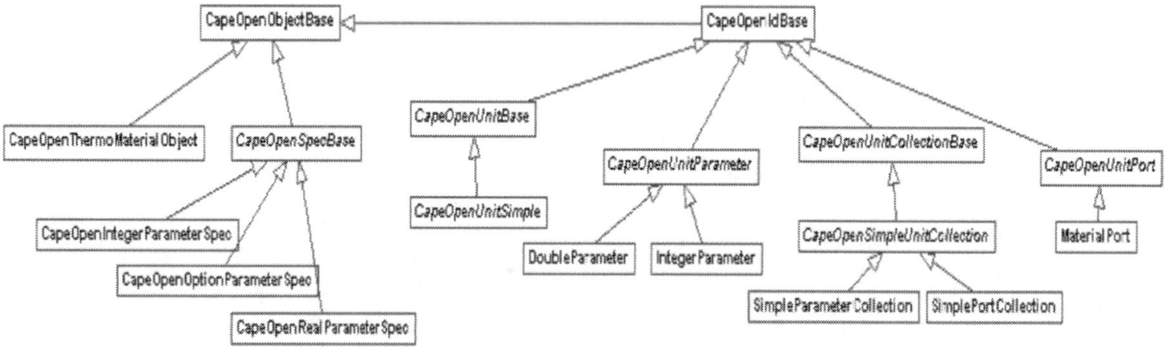


Figure 2: class based framework.

## Example of application: a dummy test component

One of the main features of the new class based framework is that the component developed by the user is a totally separated project from the infrastructure code. The connection is given by the fact that the class of the new component must inherit from the CapeOpenUnitSimple class in order to allow the process simulator to access the methods of the ICapeUnit interface and those of the interfaces which are connected to it. The user has also to "override" OnInitialize and OnCalculate methods.
In the following example, we apply the new concepts to the development of a 'dummy' unit operation ,namely a unit that does nothing, but reading input ports and transferring information to output ports. This example evidences the basic coding that a process engineer should do in .NET for preparing the development of her unit operation.

*Initialization*
The creation of ports doesn't cause many problems even in the original VB6 model, exception made for the fact that, owing to absent-mindedness, a Parameter could be inserted in the collection of Ports and vice versa. The new implementation prevents from this possible error. The creation of a Port, and its insertion in the collection, is obtained by means of the Add method, applied to the "Ports" object, which is of the PortCollection type. In this last's signature appears the Create static method, which returns a different sort of Port, a Material Port in this case, according to the CapeOpenPortType. If the user tries to insert a Parameter, the error would be pointed out at compilation time and not at runtime as it used to happen before.
The creation of Parameters is very complex in VB while with the proposed architecture the task is definitely simplified. As an example, if one would like to create a Parameter of the integer type, only the instructions reported in the following code fragment are necessary.

```
protected override void OnInitialize() {
    Name = "DummyUnit";
    Description = "Unit Dummy .NET";
    Ports.Add(CapeOpenUnitPort.Create(CapeOpenPortType.Material, "InputPort", "Input Port of Dummy",
CapeOpenPortDirection.Inlet));
    Ports.Add(CapeOpenUnitPort.Create(CapeOpenPortType.Material, "OutputPort", "Output Port of
Dummy", CapeOpenPortDirection.Outlet));
}
```

It is important to note that there is no need to know CAPE-OPEN standards details any more, that the operation now requires much less number of lines of codes than before and that it is no longer required to know the details of the ICapeIntegerParameterSpec interface.

*Calculate*

The first difference that is noticed in the new framework is that the object which represents the material fluxes is not accessible by ICapeThermoMaterialObject interface but is represented by the CapeOpenThermoMaterialObject class which has all the Properties that allows the simplified calculation of the equilibrium.

The same reasoning is also done for Ports: it is possible to access Ports not by ICapeUnitPort interface, but by CapeOpenUnitPort class (it would be just the same, in this case, to access them by MaterialPort class). The element is returned through index and the order is the same taht has been used in the creation of the OnInitialize method. It would be the same to use the GetPortByName method, here applied to the object "Ports".

The tests on direction and on type are the same, but with the difference that the used enumerations are those defined in this new project. The return of the material flux associated with the Port is obtained in any case by means of the ConnectedObject method. The cast on the returned type is a warranty of consistency among the types.

Up to now, the changes imposed by the new framework are not so relevant to justify the re-implementation of the code. The return of the values of the flux properties in the form of variables makes the difference since they are consistent with the type of quantity they represent. The same applies to the setting of these values.

Finally the CalcEquilibrium method is recalled where the possible values of the signature are represented by an enumeration. The following fragment shows the code necessary for the implementation of the dummy unit operation with the new framework. The code is particularly simple and clean.

```
protected override void OnCalculate()
{
    double zTemperature = InputPort.Temperature;
    double zPressure = InputPort.Pressure;
    double zEnthalpy = InputPort.Enthalpy;
    double zTotalFlow = InputPort.TotalFlow;
    double[] zFraction = InputPort.Fraction;

    OutputPort.Temperature = zTemperature;
    OutputPort.Pressure = zPressure;
    OutputPort.Enthalpy = zEnthalpy;
    OutputPort.TotalFlow = zTotalFlow;
    OutputPort.Fraction = zFraction;

    OutputPort.CalculateEquilibrium
       (CapeOpenFlashType.PH);
}
```

## The CapeOpenToolkit.NET

The code described above is the fundamental part of the development of a fully functional unit, but some infrastructure elements are also necessary. In this project we have developed a complete toolkit to minimize all the complementary activities. The elements of this toolkit are the following.

*Unit Wizard*

The original CapeOpen Wizard generated a complete Visual Basic 6 project. We have developed a new Wizard that targets the current Microsoft .NET Framework 2.0, generating a compatible Microsoft Visual Studio 2005 project.

Some goodies were added: (i) Load/Save of project generation execution, (ii) Typed parameters and ports, (iii) Multi language targets (C#, Visual Basic .NET) and (iv) COM Registration metadata.

*Regunitasm.EXE*

The .NET unit implementation cannot avoid the necessary COM plumbing. The new project template contains all the information needed for a registration activity, now saved at assembly level. When the generated assembly is installed in the target machine containing the process simulator, this utility can register the unit with a simple command:

    REGUNITASM <unit.dll> <Unit Type fullname>

No Windows COM registry info will be needed to perform this tedious operation.

## Conclusions

The objective of the development of the new framework v ... development of software that allows the creation of a component module to be used in process simulators in a ... e and self-evident way. The solution proposed in this paper has actually solved most of the problems which appeared in the previous models and has thoroughly reached its goals. In summary, the following items evidence the benefits and the significance of the new framework.

1. the re-writing of a component is effectively accessible also to a process engineer who has a basic knowledge of informatics;
2. most errors are pointed out during compilation, thus meaning a faster code debugging;
3. it is possible to exploit more complex data-structures, which are available in .NET;
4. a reference handbook of few pages would allow the user to access most of the functions she is interested in without concerning about CAPE-OPEN standard details;
5. the inheritance of the classes allows writing base-components which can be re-used many times and which are extendible with further inheritance.

The last point is particularly interesting. Assume, as an example, that one should develop different unit operations, all having a single inlet Port and a single outlet Port with 'some' modification of the material streams inside the unit. The way to proceed is the following. First the user creates a class, made abstract, that basically does what reported above for the dummy unit operation. After, the user will concentrate on the derived classes, one for each module to be created, each one containing a method, inserted between the returning and the setting of properties, with the implementation of the equations relating input and output streams. By doing this, there is no need to know how Ports and Streams are represented in programming language.

Currently, our effort is in porting our code base to the CAPE-OPEN 1.0 specification, thus extending the testing platform to other process simulators. So far the framework has been tested with the tester suite (13), Aspen+, PRO II and COCO/COFE. In running the tests, we have discovered many incompatibilities of 0.93-version based PMC with the PMEs considered. First tests on 1.0 are showing that there is a broader compatibility, also with PMC developed with our toolkit.

We plan to further develop the framework with the final goal of the complete hiding of the code to the final user in favor of a developer-environment oriented to the process engineer user. This environment will allow the user to write the equations describing the unit operation to be developed without taking care of how the equations are translated in effective code. This operation will be simple also for a process engineers who has no familiarity at all with informatic tools and code.

The last step of the development of the framework will be the implementation of an application developed specifically for the end-user, in which she will write the equation directly in a 'natural way'. Since it is not possible to separate informatics from engineering, there is the need to provide the final user (the process engineer) with a simplified and self-explaining development environment which doesn't require excessive notions of informatics, like Polymath or Matlab. This project has the aim to open a road towards that goal.

## Acknowlegements

Thanks to Marco Carone and Letitia Toma for helpful discussions and their contribution in toolkit creation.

## References


1. Cape open standard, Version 1.0 Documentation set, 2007, www.colan.org.
2. http://en.wikipedia.org/wiki/Microsoft_.NET_Framework.
3. http://en.wikipedia.org/wiki/Visual_Basic.
4. Parenzan M., Fermeglia M., 2007, Recent developments and implementations of CAPE-OPEN in .NET with demo, 4th CAPE-OPEN European Conference, Heidelberg, March 2007.
5. Fermeglia M., Parenzan M., 2007, CAPE-OPEN interface revisited in term of class-based framework: an implementation in .NET, ICheaP-8 conference, June 24 -27, 2007, Ischia, Italy.
6. http://en.wikipedia.org/wiki/Object_oriented_programming
7. http://en.wikipedia.org/wiki/Source_code_generation
8. http://en.wikipedia.org/wiki/Virtual_function
9. http://en.wikipedia.org/wiki/Visual_Basic_.NET
10. http://en.wikipedia.org/wiki/Com_interop
11. CO-LaN Methods & Tools Special Interest Group. *.NET Interoperability Guidelines.* CO-LaN, revision 2007.
12. CO-LaN. *Migration Support.* http://www.co-lan.org/index-9.html.
13. CO-LaN. *CO Tester Suite.* http://www.co-lan.org/index-10.html.

# 398a Unit Operation Prototypes for the Development of Cape-Open Refinery Reactors Interfaces

**Ignasi Palou-Rivera**, Model Development Team, BP Products North America, Inc., 150 West Warrenwille Road, Mail code E-1B, NAPERVILLE, IL 60563-8460 and Jeffrey Pribich, SAIC, Lisle, IL 60632.

This presentation will focus in the experiences of developing a "refinery" unit operation as a prototype for the development and validation of interfaces as part of COLaN's Refinery reactors SIG. In the context of Co-LaN's Refinery Reactors SIG, we define a "Refinery" Unit Operation as interacting with compounds defined by modifiable (petroleum) properties. We can propose a simple classification of "refinery" unit operations into three types: reactors, property calculators, and blenders. Form an architectural perspective these are only two types of units: those with a single feed whose output has the properties of the combined feed, and those with a single feed whose output properties are set by the unit in any determined way. Therefore two prototype models are needed to test and develop the necessary interfaces: property calculator, and blender. The chosen prototype units are a simple octane calculator (RON is calculated from the feed stream and written in the product stream), and a blender where a blended oil stream is determined from multiple defined feed streams. These prototypes have been developed following this sequence of events:

1) Architecture definition: number of feed and product streams, parameters, etc

2) Unit operation structure is built using CAPE-OPEN Unit Operation VB Wizard.

3) Specialized Refinery reactors CO interfaces are hand-coded.

4) Purpose calculation code is added (RON calculation, crude mixing equations, etc)

5) Unit is tested in several CO-compliant PMEs: ProII, Unisim, Hysys, COCO

## Use of CAPE-OPEN Standard in US-UK Collaboration on Virtual Plant Simulation

**Stephen E. Zitney**
Collaboratory for Process & Dynamic Systems Research
National Energy Technology Laboratory, Morgantown, WV 26507-0880
stephen.zitney@netl.doe.gov www.netl.doe.gov

AIChE 2007 Annual Meeting, November 4-9, Salt Lake City, UT

Under the auspices of a US-UK Memorandum of Understanding and Implementing Agreement for fossil energy R&D (http://us-uk.fossil.energy.gov/), the US Department of Energy's (DOE) National Energy Technology Laboratory (NETL) and the UK Department of Trade and Industry (DTI) have recently completed a three-year collaboration on virtual plant modeling and simulation technology for advanced fossil-energy power generation systems. The R&D collaboration was aimed at taking full advantage of the synergies between NETL's ongoing Advanced Process Engineering Co-Simulator (APECS) project and the UK's three-year Virtual Plant Demonstration Model (VPDM) project. The key objective of this collaboration has been the development of compatible, open standards-based US and UK technology for process/equipment co-simulation. To achieve plug-and-play model interoperability, the collaboration leveraged the process-industry CAPE-OPEN (CO) software standard which is managed and disseminated by the CO Laboratories Network (www.colan.org).

We highlight here the use of the CO interfaces for facilitating the interoperability of power plant equipment models between the APECS and VPDM software platforms. Special emphasis will be focused on the use of the CO steady-state unit operation interfaces to integrate process simulation with a hierarchy of equipment models ranging from computational fluid dynamics (CFD) models to fast reduced-order models (ROMs) based on CFD results. Virtual plant simulations based on CO compliant process/CFD co-simulations will accelerate the design of next-generation plants to operate with unprecedented efficiency and near-zero emissions, while operating profitably amid cost fluctuations for raw materials, finished products, and energy.

### APECS

NETL's **Advanced Process Engineering Co-Simulator (APECS)** system is an integrated software suite that combines process simulation, high-fidelity CFD, immersive and interactive 3D plant walk-through virtual engineering, and advanced analysis capabilities including case studies, sensitivity analysis, stochastic simulation for risk/uncertainty analysis, and multi-objective optimization. To facilitate effective technology transfer, the APECS system was designed using commercial process simulation and CFD software with model interoperability provided by the CAPE-OPEN interfaces for unit operations, physical properties, and reaction kinetics.

Using APECS, design engineers can effectively integrate and efficiently solve and analyze co-simulations combining the CO-compliant, steady-state process simulators such as Aspen Plus® (www.aspentech.com), Aspen HYSYS®, and gPROMS® (www.psenterprise.com) with CO-compliant process equipment models, including CFD models based on FLUENT® (www.fluent.com), the world's leading software for detailed flow analysis of process equipment. APECS also offers fast ROMs based on pre-computed CFD solutions over a range of parameter values. For example, the APECS system currently provides for automatically generating and using ROMs based on multiple linear regression and neural networks. Future ROM solvers will include principal component analysis (PCA) and proper orthogonal decomposition (POD). The CAPE-OPEN COM/CORBA bridge implementation in APECS allows process simulations running under the Windows operating system to use equipment models running locally/remotely and serially/in parallel on Linux clusters and/or supercomputers.

Advanced 2D and 3D visualization tools, including Paraview (www.paraview.org), available in APECS enable design engineers to display, within the process simulator, the results of a process/CFD co-simulation including contours of velocity, temperature, pressure, and species mass fractions for specified surfaces in the equipment models. Ongoing collal ve efforts aimed at integrating APECS with VE-

Suite (www.vesuite.org), an open-source virtual engineering software toolkit, will enable users to explore co-simulation results in a context-based, user-centered interface including walking through a 3D representation of the power plant.

The APECS system also provides a wide variety of powerful analysis tools for optimizing overall power plant performance. Optimization is used for maximizing an objective function, including plant efficiency, energy production, and process economics. For process optimization in the face of uncertain variables and multiple and some time conflicting objectives, APECS offers stochastic modeling and multi-objective optimization capabilities developed to comply with the CO software standard.

In March, 2005, U.S. Secretary of Energy, Samuel Bodman, announced that Ansys/Fluent (www.ansys.com) and Reaction Engineering International (www.reaction-eng.com) were awarded APECS projects, respectively, as part of a portfolio of 32 clean coal research projects. These phase-II APECS software development projects are well underway and have been augmented recently with a portfolio of APECS R&D projects launched under the auspices of NETL's Collaboratory for Process & Dynamic Systems Research organized in partnership with Carnegie Mellon University, the University of Pittsburgh, and West Virginia University.

At NETL, system analysts are applying APECS to a wide variety of advanced power generation systems, ranging from small fuel cell systems to commercial-scale power plants. Using APECS, the overall performance of solid oxide fuel cell (SOFC) auxiliary power units (APUs) system modeled using Aspen Plus are optimized with respect to the local fluid flow, heat and mass transfer, electrochemical reactions, current transport, and potential field in the SOFC simulated using detailed, three-dimensional, steady-state FLUENT CFD models. The process/CFD co-simulations are performed over a range of fuel cell currents to generate a voltage-current curve and analyze the effect of current on fuel utilization, power density, and overall system efficiency.

In collaboration with cycle engineers at Alstom Power (www.power.alstom.com), APECS co-simulations have been developed for a conventional 30 MWe pulverized coal-fired (PC) steam plant for municipal electricity generation and an advanced 250 MW, natural gas-fired, combined cycle (NGCC) power plant. In the PC co-simulation, an Aspen Plus process design specification is used to adjust a FLUENT CFD model parameter for the boiler damper position (bypass resistance) to maintain a specified steam temperature over a range of loads, from the load at the maximum continuous rating to a control load, below which the boiler cannot sustain the required turbine inlet temperatures. For the NGCC co-simulation, an Aspen Plus process design specification is used to manipulate designated control parameters for the FLUENT CFD model of the heat recovery steam generator (HRSG) so that a specified superheat steam temperature is maintained for various load points over the range from 100% to 50% gas turbine load.

At NETL, research engineers are also developing APECS co-simulations to analyze potential FutureGen plant configurations. The DOE's 10-year, FutureGen Research Initiative is aimed at creating the world's first coal-based, near-zero emissions electricity and hydrogen production power plant with carbon capture and storage (www.futuregenalliance.org). In a recent demonstration case, the FutureGen co-simulation combines a plant-wide Aspen Plus simulation with two FLUENT CFD-based equipment models, one for the entrained-flow gasifier where fluid dynamics strongly affect syngas quality and carbon conversion and one for the gas turbine combustor where the blending of air and fuel is at the heart of gas turbine combustor performance and efficiency. Using APECS, Aspen Plus controls the co-simulation and automatically executes the gasifier and combustor CFD models as needed to converge the tail gas recycle loop and a design specification on the gas turbine inlet temperature. The design specification is met by manipulating the synthesis gas split between power production and hydrogen production.

The co-simulation applications described here illustrate how APECS is helping engineers optimize the coupled fluid flow, heat and mass transfer, and related phenomena that drive overall system performance. By combining process simulation and CFD software, together with advanced visualization and high-performance computing, APECS provides the high-fidelity solution and analysis capabilities required to achieve the aggressive integration, environmental, and performance goals for advanced power generation systems such as the DOE's FutureGen plant.

**VPDM**

The principal objective of the UK's **Virtual Plant Demonstration Model (VPDM)** project was to develop a software framework for accurate power plant modelling. The project had participants from UK companies and academia, and received support from the DTI Carbon Abatement Technologies Programme. The partners included: ALSTOM Power Ltd., Engineous Software Ltd., Ansys Fluent Europe Ltd., RWE npower plc, K-S Tech Ltd., Doosan Babcock Energy Ltd., Process Systems Enterprise Ltd. (PSE), and the University of Ulster. The concept of the VPDM is that it is able to utilize the most accurate plant equipment models available, many of which are proprietary and residing on various hardware platforms at different locations. The VPDM team leveraged NETL's APECS technology to integrate high-fidelity FLUENT CFD equipment models into overall power plant simulations developed with PSE's gPROMS simulator. The project also exploited the common standards for interfacing based on the CAPE-OPEN standard. In the development of the VPDM, the following CO-compliant process simulators were tested: gPROMS from PSE, Aspen Plus from Aspen Technology, and ECLIPSE from the University of Ulster.

Another key feature of the VPDM process modeling environment is that it allows secure web interfacing and data exchange with individual heterogeneous component models located remotely at partner sites. Thus, companies can allow access to their in-house equipment models without compromising the intellectual property contained within the models. The VPDM framework and the availability of such models will provide the opportunity to investigate, through simulation studies, accurate novel power plant solutions which are capable of producing power in a reliable, cost-effective and environmentally friendly manner. The proposed VPDM framework uses the web technology contained within FIPER from Engineous (www.engineous.com) to provide a secure client-server communications layer between partner companies over the Internet. Some special software developed by PSE and Engineous, as part of the VPDM program, enables FIPER to communicate with CO-compliant process simulators and CO-compliant equipment models.

The initial target application for the VPDM was the simulation of a 500 MW conventional coal-fired power plant. Using CO-based process/CFD co-simulation technology in NETL's APECS system, the VPDM team coupled a detailed FLUENT CFD furnace simulation with a steady-state gPROMS process simulation of the power plant. As a demonstration of the VPDM's versatility, it was also successfully applied to two new fossil power plant technologies which both include carbon capture.

**US-UK Collaboration Tasks**

Joint APECS/VPDM work began in January 2005 on four key collaborative tasks which are outlined below.

*Collaboration Task 1: Using CAPE-OPEN Interface for Interoperability of Equipment Component Models and Virtual Power Plant Simulation Frameworks*

The objective of this collaboration task was the exchange of information and experience in the development and application of CO-compliant equipment models based on the FLUENT CFD code for use in the UK and US projects. The following benefits were achieved:

- Developed plug-and-play capability, between US equipment component models and the VPDM framework, as well as between UK equipment models and NETL's APECS simulation frameworks.
- Placed initial emphasis on using FLUENT CFD models within both software environments.
- Focused on using the process industry standard CO interfaces.

*Collaboration Task 2: Using Reduced-Order Component Models to Enhance Speed of CFD models for Virtual Power Plant Simulation Frameworks*

A potential barrier to the widespread use of process/CFD co-simulation is that the integration of high-fidelity equipment models may lead to unacceptable co-simulation turnaround times, especially for cases in which one or more CFD models are embedded in the iterative flowsheet solution process. One promising solution is the use of reduced-order models (ROMs) that approximate the CFD-based equipment simulations, while keeping the computational cost manageable. The key benefits of this collaboration task were the following:

- Allowed US DOE/NETL and UK (PSE) researchers to collaborate in their evaluation of order reduction techniques for application to power plant simulation.
- Collaborated to devise strategies for implementing order reduction techniques in respective UK and US programs.
- Exchanged information on accuracy and speed of ROMs as implemented in the simulators.

*Collaboration Task 3: Internet Security and Communication*

The fundamental aim of both the UK's and the US's projects is to obtain maximum accuracy from power plant simulation models by linking together existing plant component models that have benefited from many years of development, validation, and refinement. These models contain propriety information and each model owner will want to operate their model from their own site, where they have confidence that their know-how is secure. The main benefit of this collaboration task was the exchange of information to determine the specification of security requirements suitable for users of the simulation frameworks.

*Collaboration Task 4: CAPE-OPEN Dynamic Simulation*

Both the US and UK virtual power plant simulation programs are planning to pursue dynamic simulation, especially for use in process control. The process industry-standard CO interfaces currently support steady-state modelling only, but a standard for dynamic modelling is under development. The benefits of this collaboration task were as follows:

- Reviewed current status of CO dynamic simulation standard.
- Identified the dynamic modelling requirements of the APECS and VPDM projects.
- Worked with the CO developers to ensure that the CO interfaces for dynamic simulation are developed in a timely manner and meet the agreed requirements of the VPDM and APECS projects.

**Collaboration Benefits**

The US-UK collaboration has worked well; in particular the joint adoption of the CAPE-OPEN software standard will have benefits on both sides of the Atlantic and allow interoperability between equipment models in the US APECS project and UK VPDM project. The US project is currently focusing on using FLUENT to create appropriate CFD-based models for advanced power plants and these models will be run with the Aspen Plus process simulator. The success of this collaboration task means that it will be possible to use these CFD models with the VPDM framework. Similarly, the UK FLUENT based CFD models can be used in the US APECS system. The exchange of ideas and information on Internet security and protecting IP has resulted in an organization external to the VPDM project approving of the techniques used in the VPDM; this gives confidence for the future adoption of the VPDM principals beyond the VPDM project.

By leveraging the US APECS project, UK VPDM project, and the international process-industry CAPE-OPEN software standardization effort, the US-UK collaboration on virtual power plant simulation achieved its main goal which was to develop compatible process/CFD co-simulation software platforms for application to advanced fossil energy systems. The key benefits to US and UK fossil energy industries include:

- Helps engineers to better understand and visualize the fluid flow behavior that impacts process design and operation
- Considers detailed equipment models in the context of plant-wide simulations, with recycle loops, heat integration, and water management
- Enables rigorous analysis and optimization of entire plants with respect to CFD-related equipment model parameters
- Eliminates potential for suboptimal designs by using same physical properties and reaction kinetics in the underlying equipment and process models
- Speeds technology development by reducing pilot/demo-scale facility design time and operating campaigns
- Offers opportunities to achieve the aggressive environmental, performance, and economic goals for high-efficiency, zero-emission power plants

Continued US-UK collaboration on virtual power simulation will accelerate the development of competitive power plant solutions and ultimately zero-emission technologies with significantly reduced development costs and technical risk.

Extended Abstract submitted to 4th US Cape-Open Conference, AIChE Annual Meeting, Salt Lake City, November 4-9, 2007

# Advanced Process Engineering Co-Simulation using CFD-based Reduced Order Models

**Yi-dong Lang[1,2], Lorenz T. Biegler[1,2]**
**Sorin Munteanu[3], Jens I. Madsen[4]**
**Stephen E. Zitney[2]**

[1]Department of Chemical Engineering, Carnegie Mellon University, Pittsburgh, PA
[2]Collaboratory for Process & Dynamic Systems Research, National Energy Technology Laboratory, Morgantown, WV
[3]Ansys Inc., Lebanon, NH
[4]Ansys Inc., Morgantown, WV

## 1. Preface

The process and energy industries face the challenge of designing the next generation of plants to operate with unprecedented efficiency and near-zero emissions, while performing profitably amid fluctuations in costs for raw materials, finished products, and energy. To achieve these targets, the designers of future plants are increasingly relying upon modeling and simulation to create virtual plants that allow them to evaluate design concepts without the expense of pilot-scale and demonstration facilities. Two of the more commonly used simulation tools include process simulators for describing the entire plant as a network of simplified equipment models and computational fluid dynamic (CFD) packages for modeling an isolated equipment item in great detail by accounting for complex thermal and fluid flow phenomena. The Advanced Process Engineering Co-Simulator (APECS) sponsored by the U.S. Department of Energy's (DOE) National Energy Technology Laboratory (NETL) has been developed to combine process simulation software with CFD-based equipment simulation software so that design engineers can analyze and optimize the coupled fluid flow, heat and mass transfer, and chemical reactions that drive overall plant performance (Zitney et al., 2006). The process/CFD software integration was accomplished using the process-industry standard CAPE-OPEN interfaces.

A potential barrier to the widespread use of process/CFD co-simulation is that the integration of high-fidelity equipment models may lead to unacceptable co-simulation turnaround times, especially for cases in which one or more CFD models are embedded in the iterative flowsheet solution process. One promising solution is the use of reduced-order models (ROMs) that approximate the CFD-based equipment simulations, while keeping the computational cost manageable. ROMs are equipment models that contain only relevant states or a reduced number of irrelevant states. The advantage is dramatically increased simulation speeds and lower memory requirements.

To obtain ROMs for equipment items with arbitrary geometries, we propose a novel state-space ROM strategy. In contrast to the popular order reduction method of proper orthogonal decomposition (POD), this work focuses on a principal component analysis (PCA) approach. First, the statistical Latin hypercube sampling (LHS) technique is used for experimental design to determine a distribution of plausible sets of operation and boundary conditions for the CFD model. Second, using off-line CFD simulations with the designed cases, we obtain a database consisting of snapshots of the flow-field variables (state variables). Third, selecting the snapshots of one state variable of interest, we formulate a snapshot matrix and implement PCA on it to derive the ranked principal components (PC) and the coordinates of the data elements in the transformed vector space (scores). Fourth, using a neural network formulation, we train a mapping from the inputs of CFD model into the scores. Finally, we assemble the ranked PC and mapping score to develop the ROM.

A case study is presented based on a 2-D FLUENT CFD model of a gas turbine combustor with ten input variables. The PCA-based ROM can be executed in seconds leading to CPU savings of two to three orders of magnitude over the high-fidelity CFD model. Replacing the CFD model in the process flowsheet with the ROM through the standard CAPE-OPEN unit operation interface therefore allows APECS process/CFD co-simulations to be solved more efficiently and effectively.

## 2. Development of ROM for a Gas Turbine Combustor

At NETL, system analysts are applying APECS to reduce the time, cost, and technical risk of developing high-efficiency, near-zero emissions power plants such as the coal-fired, gasification-based plant in the $1 billion, 10-year DOE FutureGen R&D Initiative. To build a comprehensive process/CFD co-simulation of the FutureGen power plant, NETL is developing high-fidelity CFD models for process units in which fluid flow, heat and mass transfer, and chemical reactions impact performance of the unit as well as the overall plant. Initial efforts have focused on the development and use of CFD models for the gasifier and gas turbine combustor. Gasifier fluid dynamics strongly affect syngas quality and carbon conversion. Similarly, the blending of air and fuel is at the heart of gas turbine combustor performance and efficiency.

A 2D FLUENT CFD model of the gas turbine combustor is used to test the ROM methodology developed in this study. The general method of Latin Hypercube Sampling (LHS) is used to generate an experimental set of values for the input variables defined as operating conditions and boundary conditions for the CFD model. First, statistical evaluations are used to detect correlations between variables. For the combustor model, we reduced the original ten inputs to nine independent inputs by creating a new variable, mole ratio of oxidant ($O_2$) to fuels (CO, $H_2$, and $CH_4$) for use in maintaining a feasible combustion process. Then based on the possible range of input variables encountered during iterative convergence of the process/CFD co-simulation, we determined the range of inputs within which the developed ROM is valid. In this study, 128 experimental samples are designed and 128 corresponding FLUENT cases are solved. We use the graphic facilities of Matlab to plot the results and compare the differences of the temperature fields from the CFD solution and the ROM.

With the database obtained from FLUENT, we implement PCA using Singular Value Decomposition (SVD) in Matlab. We showed that PCA with rank 35 can capture 99.999% of the energy in the 128 cases, and so we used 35 principal components to develop the ROM. The PCA decomposes the database into two parts; the principal components are new coordinates that are fixed while the scores are the projection of the database into the coordinates, respectively, determined by the inputs sets. In other words, if we establish the relationship between the scores and inputs, we can relate the temperature field to the inputs combined with the coordinates. This is our goal with the derived ROM. In addition, we use a neural network to regress the functional relationship of the scores from the PCA to the inputs of the FLUENT cases. Mathematica [illegible] eaking, this implements the mapping from the inputs to the scores.

Once the ROM was derived for the gas turbine combustor, we tested and analyzed its accuracy and validity. Here we make two observations.

1. The ROM can reproduce the temperature field and the outlet temperature from the combustor accurately at the 128 experimental points. The error of the outlet temperature is less than $10^{-7}$.
2. We observed that in the neighborhood of the existing data points, if the radius is small, the derived ROM provides a good approximation in comparing with the solution with FLUENT at the same input point. This phenomenon predicts that if the density of the existing points in the input space is big enough, then the derived ROM based on that database will be good for that input space. Also we observed that the database we obtained with 128 points is still not big enough to be a good ROM for the entire input range.

## *Gas-Turbine Combustor*

In the FutureGen process model flowsheet, the combustor is an important reactor that combusts synthesis gas to produce a high temperature exhaust that gas stream, which drives the turbine. The fuel is originally produced by a gasifier followed by a series of pre-treatment procedures such as water gas shift reaction and $CO_2$ quench. The syngas is premixed with compressed atmospheric air to form the combustor input stream. The variables included in this input stream are mass flow rate, temperature and mass fractions of eight components (seven out of eight are independent). The components are $CO_2$, CO, $O_2$, $H_2$, $H_2O$, $CH_4$, Ar and $N_2$. Among these, only CO, $H_2$, $CH_4$ and $O_2$ are involved in the combustion reactions.

In this study, we use a FLUENT CFD model of the combustor. This is a turbulent, lean-premixed, swirl-stabilized combustor based on the NETL research combustor described by Sidwell et al. (2005). The CFD model is scaled up to represent a single combustor can in a 250MW, 16 combustor can gas turbine used for the FutureGen power plant. The 2D axisymmetric combustor geometry with nearly 6000 computational cells is used to solve the CFD model.

The flame is anchored by careful scaling and design of the combustor, so that the nozzle velocity of the NETL research combustor is matched. In the FLUENT model of the combustor, the finite rate/eddy dissipation model is selected. The reaction rate is defined by taking the minimum of the chemical reaction rate and the turbulent mixing rate. The global reactions in the combustor are:

$$2CO + O_2 \rightarrow 2CO_2$$
$$2H_2 + O_2 \rightarrow 2H_2O$$
$$CH_4 + 2O_2 \rightarrow CO_2 + 2H_2O$$

With the CFD model, we can change the input variables (in terms of inlet stream information in Aspen Plus/APECS) that represent operating conditions and boundary conditions in FLUENT, to obtain corresponding state fields and output stream(s). In this study, we consider one state, temperature, only. Nevertheless, the methodology of deriving the ROM for temperature can be applied directly to any other state without any restriction.

## *Experimental Design*

Because the combustor model has multiple inputs, we need to minimize the number and size of the database used for development of the ROM and maximize the information covered for the database. Here it is necessary to invoke the techniques of experimental design to obtain a designed set of inputs that cover the ranges of the inputs. In this study; we chose the Latin Hypercube Sampling (LHS) method with uniform probability distribution for the nine independent input variables.

Before we implement LHS, we reformulate the input space first to make all components or coordinates spanning the input space non-correlating, i.e. independent of each other.

By inspecting the chemical behaviors of the combustor and the streams, we found the molar fractions of CO and H2 are correlated and dominated by the ratio of CO/H2 controlled by the gasifier. In addition, to maintain proper combustion the molar ratio of oxidant to fuel must be constrained. According to the flowsheet, we assume that there are two streams, Fuel and Air, mixing together and forming one stream entering the combustor. With some assumptions for simplicity, we finally ended up with the following nine independent input variables: operation pressure, inlet temperature, mass flowrate, inlet mole fractions of CO, CO2, H2O, and CH4, the molar ratio of CO/H2 in fuel stream, and the molar ratio of (CO+H2+CH4)/O2 in the fuel.

The experimental design is implemented using Latin Hypercube Sampling to produce a set of experimental independents with parameters:

- No. of independents without correlation = 9
- No. of experimental samplings = 128
- Probability distribution = Uniform

LHS gives a designed set of independent variables with 128 cases. In each case, there are sampling values for the nine variables within the ranges. Since some of these independent inputs cannot be prescribed as boundary conditions directly in FLUENT, we map them into corresponding FLUENT inputs for each case.

## *Building Database*

It is straightforward to perform a batch of FLUENT simulations for the designed sampling set. For each group of the inputs in the designed sampling set, a case is established by solving the FLUENT CFD model first, then extracting the temperature field and integrated (mass average weighted) outlet temperature. When all 128 cases are solved, we obtain a database matrix with 5996 (number of mesh points) rows and 128 columns, each column of the matrix corresponding to the temperature field of one case. Also, we obtain a vector where each entry represents the outlet temperature of corresponding case.

## *PCA applied to the database*

The established database is expressed as the snapshot matrix $Y_T$. Implementing PCA on the matrix, $Y_T$ we obtain

$$Y_T = U\Lambda V^T .$$

By choosing the ratio of captured energy as 99.999%, the rank is 35. Here we use the first 35 columns of $U$ and first 35 rows of $V^T$ to formulate the PCA approximation of the database matrix, i.e.

$$Y_T \approx U_r \Lambda_r V_r^T .$$

Defining the score as:

$$\Omega \approx \Lambda_r V_r^T \in \Re^{35x128}$$

leads to:

$$Y_T \approx U_r \Omega .$$

It is known that only the score corresponds to the input that determines the database. In the next section we will discuss mapping the input set to the score with the neural network method.

## *Regression using Neural Network*

Principal Component Analysis on the temperature snapshot matrix $Y_T \in \Re^{2601\times128}$, with 99.999% energy captured, produced 35 corresponding ranked principal components

$$U_r \in \Re^{2601\times35}$$

and scores

$$\Omega \in \Re^{35\times128} .$$

To use one neural network for the ROM to accommodate the scores and the outlet temperature, we add outlet temperatures in 128 cases into the last row of the matrix $\Omega$, we get

$$\Omega' \in \Re^{36\times128} ,.$$

In order to build relationship between the independent experimental design $Indep \in \Re^{9\times128}$ and the ranked scores and outlet temperature in $\Omega' \in \Re^{36\times128}$, a back-propagation neural network is created with 800 nodes in the hidden layer and 36 nodes in output layer respectively. It is worth noting a fundamental property of neural networks, i.e., "Multi-layer networks are quite powerful. For instance, a network of two layers, where the first layer is sigmoid layer and the second layer is linear, can be trained to approximate any function (with a finite number of discontinuities) arbitrarily well." --- H. Demuth et al (2005).

The mapping can be expressed as

$$\Omega' = F_2(LW \cdot (F_1(IW \cdot Indep + b\{1\}) + b\{2\}) ,$$

where $F1$ and $F2$ are transfer functions for layer 1 and 2, respectively. $F1$ is a Tan- Sigmoid and $F2$ is just linear. With $\Omega' \in \Re^{36\times128}$ as the target and $Indep \in \Re^{9\times128}$ as the input, we train the neuron network with $MSE$ (Mean Square Error) <= $10^{-16}$ to get the parameters

$$IW\{1,1\} \in \Re^{800\times9} , b\{1\} \in \Re^{800\times1} ,$$

$$LW\{2,1\} \in \Re^{36\times 800}, b\{2\} \in \Re^{36\times 1},$$

simultaneously.

Using obtained parameters of the network we build a new network for network simulation, i.e. for a given input vector, *u,* calculate the corresponding output of the network, the score,

$$\omega(u) = F_2(LW \cdot (F_1(IW \cdot u + b\{1\}) + b\{2\}) \in \Re^{36}$$

as well as outlet temperature. Then, getting back the principal component, we formulate the Reduced Order Model for the temperature field in the combustor and outlet temperature as,

$$\begin{bmatrix} y_T \\ T_{ROM}^{out} \end{bmatrix} = \begin{bmatrix} U_r & 0 \\ 0 & 1 \end{bmatrix} \omega(u)$$

## *ROM evaluation*

In order to evaluate the ROM thoroughly, two different comparisons are made. First, we test the predictive accuracy of the ROM at the LHS design points. Second, we evaluate the approximation quality of the ROM at the random points within the designed range in the nine-dimensional input space.

### A. Evaluation at the designed data points

We now can have three data sets obtained by FLUENT simulation, PCA approximation, and the derived ROM. Each set contains 128 cases with different operating and boundary conditions. For each FLUENT simulation, we have one temperature field which consists of 2601 nodes, as well as the value of the outlet temperature averaged over the mass flow. All of the temperature files of 128 cases construct a 2601 x 128 snapshot matrix. With PCA on the matrix, we obtained ranked Principal components and corresponding scores with rank of 35 with 99.999% energy captured. Then we developed a ROM for PCA approximation along with outlet temperatures included in the ROM. We denote the resulting FLUENT temperature fields as true values of the combustor model, written as $TF_{true}$. Similarly $TF_{PCA}$ represents the fields approximated by the ranked PCA and $TF_{ROM}$ represents the field calculated with the derived ROM. Let $T_{true}^{out}$ and $T_{ROM}^{out}$ denote the outlet temperatures integrated by FLUENT and calculated by the ROM, respectively.

We plot the three temperature fields in filled and unfilled contours separately using data in $TF_{true}(i)$, $TF_{PCA}(i)$ and $TF_{ROM}(i)$, respectively, corresponding to the case *i*. Then we compare these plots one-by-one and conclude that in some of them there is a little observable distortion between $TF_{true}$ and $TF_{PCA}$, even though most of them have no distinguishable differences. It is possible to compare the outlet temperatures obtained by FLUENT and the ROM quantitatively. The magnitude of the differences is $10^{-8}$ to $10^{-9}$. Therefore, we concluded that the ROM is an accurate approximation of the FLUENT model solutions at all of the designed independent data points.

Qualitatively,

$$TF_{ROM}(i) = TF_{PCA}(i) \approx TF_{true}(i), \quad i = 1,128$$

Quantitatively,

$$T^{out}_{ROM}(j) = T^{out}_{true}(j), \quad j = 1,128$$

### B. Evaluation at random points

We test the ROM at a set of 128 random points in the nine-dimensional parameter space. First we pick up a pair of random integer number between 1 and 128, e.g. *ip_1* and *ip_2*, as the indices of cases in the database. Second we use a random number between 0 and 1, e.g. *w1* as the weight. Then input of the test point is: *Indep_t = Indep(ip_1)*w1 + Indep(ip_2)*(1-w1)* .With calculated input, *Indep_t* of the random test point, we run the FLUENT model and the ROM separately and compare the temperature field and the outlet temperature.

45 random test points are implemented and investigated which shows that the relative error of the outlet temperature

$$err(k) = \frac{|T^{out}_{ROM}(k) - T^{out}_{true}(k)|}{T^{out}_{ROM}(k)} \leq 3\%, \quad k = 1, 45$$

Not too surprisingly, the temperature fields are approximated with less error when the random point is in the proximity of a LHS design point.

The evaluation of the developed ROM reveals that the accuracy of the ROM depends strongly on the density of data points in the input space, determined by the experimental design. In the present case study, 128 data points proved sufficient to approximate the model response with acceptable engineering accuracy. The accuracy can be improved further by:

a. Reducing the dimension of the input space;
b. Narrowing the ranges of the input variables;
c. Increasing number of datasets.
d. Exploring more suitable techniques of experimental design

## 3. Summary

This case study is based on reduced order modeling of a key process unit, the gas turbine combustor in a FutureGen flowsheet. Here we show that a ROM methodology based on PCA instead of POD is simple, effective and efficient. It removes the burden of handling PDE equations directly and takes advantage of well-developed-and-tested CFD software, e.g. FLUENT. Combined with techniques of experimental design, the developed ROM can achieve reasonable accuracy with relatively small databases. For multi-input cases, the number of cases and the input variables for each case should be carefully designed. The case study also shows that the accuracy of ROM approximation at experimental points is very high. Moreover, for unknown points in the input space, we can achieve acceptable accuracy for the ROM as long as the density of the known points is large enough and well-distributed.

It is worth noting that there is significant CPU time reduction due to replacing the CFD model with the ROM. This is especially true for some CFD models that need several hours to converge, as the ROM needs only a few CPU seconds to give the approximate solution.

Finally, it is obvious that the PCA based methodology of developing ROMs for CFD models will accelerate process simulation in APECS significantly. Using the process-industry standard CAPE-OPEN interfaces and replacing CFD model with its ROM derived based on the methodology developed in this work, APECS will be more efficient and more effective.

# CAPE OPEN Modules for the Process Sustainability Prediction Framework-Description and Applications

Maurizio Fermeglia [a,b,*], Gennaro Longo [a], Letitia Toma [a]

[a] *International Centre for Science and High Technology, ICS-UNIDO, Trieste, Italy*
[b] *Molecular Simulation Engineering Laboratory, Department of Chemical Engineering, University of Trieste, Trieste, Italy*

## Abstract

Sustainability in chemical production is a basic social, economical and engineering issue. The sustainability of a given production process can be examined using different methodologies based on process and plant data. It is advisable to introduce the concept of sustainability at design stage. In other words, the algorithms for measuring the sustainability of a given process should be included directly into commercial process simulators, so that chemical engineers could base their design not only on technical and economical, but also on environmental issues. This way it could be possible to incorporate environmental indexes into an objective function and, consequently, minimize the process impact by adjusting the design and operative conditions.

The software developed should be based on international standards in order to be compatible to commercial process simulators. For this reason, the Process Sustainability Prediction (PSP) Framework was developed using the CAPE OPEN(CO) standard. In this work, the PSP Framework structure its modules and applicability through a series of examples are presented.

The framework basis is the definition of some indicators that could summarize the sustainability of a given process. One-dimensional indicators (1D) and three-dimensional indicators (3D) have been used. For the 3D indicators, the methodology proposed by Martins (Martins *et al.*, 2006) has been implemented. This methodology takes into account the material intensity, the energy intensity, the potential chemical risk and the potential chemical impact. The indicators' calculation is performed in the PSP Framework, starting form a toxicological database and from the material and energy balances, which are calculated by the process simulator and transferred to the algorithm through the CO modules. The PSP Framework contains also a 1D indicator methodology, based on the Waste Reduction (WAR) Algorithm. The methodology considers the engineering and toxicological data and calculates the impact through environmental indexes. The PSP Framework consists of four CO Unit Operation Modules interfaced with process simulators, a relational database (DB) containing toxicological and risk information, and a number of software tools.

The details regarding the development of the CO Modules and the description of the whole framework are presented in the paper. The PSP Framework has been tested using different process simulators: COCO/COFE, PROII and Aspen Plus. Representative examples are reported for methodology validation.

* Correspondent author
mauf@dicamp.units.it , Tel.: +390405583438, Fax: +39040569823

## 1. Introduction

Industrial sustainable development has been defined as the creation of goods and services using processes and systems that are non-polluting, conserving energy and natural resources, economically viable, safe and healthy for employees, communities and consumers; and socially and creatively rewarding for all working people (Veleva and Ellenbecker, 2001). In practical terms, industrial sustainability means employing technologies and know-how to use less material and energy, maximizing renewable resources as inputs, minimizing generation of pollutants or harmful waste during product manufacturing and use, and producing recyclable or biodegradable products (Jenk *et al.*, 2004).

The present work whises to introduce the concept of sustainable development during the design step of a process and to offer a valid tool for sustainability evaluation.

The 3D indicators' methodology, proposed by Martins (Martins *et al.*, 2006), is used in this framework as the first evaluation of the global sustainability of a chemical process. Four 3D indicators have been chosen: the material intensity (MI), the energy intensity (EI), the potential chemical risk (PCR) and the potential environmental impact (PEI). The choice of these indicators has been made taking into account the relevant data from a chemical plant and the three basic aspects of sustainable development (environment, economy and society).

The next step in the PSP Framework is represented by the 2D and 1D indicators' implementation. The environmental impact evaluation is made through the WAR Algorithm, which represents a 1D indicator. The environmental impact indexes' calculation is a key element in the WAR Algorithm. Four indexes are characteristic for the WAR Algorithm: the total rate of impact output(Iout), the total impact output per mass of products(Iout_mp), the total rate of impact generation(Igen) and the total impact generated per mass of product(Igen_mp) (Cabezas *et al.*, 1997).

The 3D and 1D indicators calculated are used, in the process design stage, in a comparative manner to choose the most environmentally friendly design from different alternative applications.

Details regarding the 3D and 1D indicators are presented in the next section.

## 2. Development

A general schema of the PSP Framework development is given in Figure 1.

There are two starting points for calculating the sustainability indicators: the process simulator and the database containing toxicity and risk phrases data. To be effective, the proposed software (PSP Framework) should be based on international standards, in order to be compatible to any commercial simulator. For this reason the PSP Framework has been developed using the CO standard.

Two CO Modules have been developed for 3D indicators and two CO Modules for 1D indicator. All the modules have been created as unit operations which, inserted in the process flow-sheet, are able to extract all the necessary data requested for sustainability indicators calculations. Data extracted with the CO Modules are processed in PSP software. The graphical user interface allows the choice of 3D and/or 1D indicator(s).

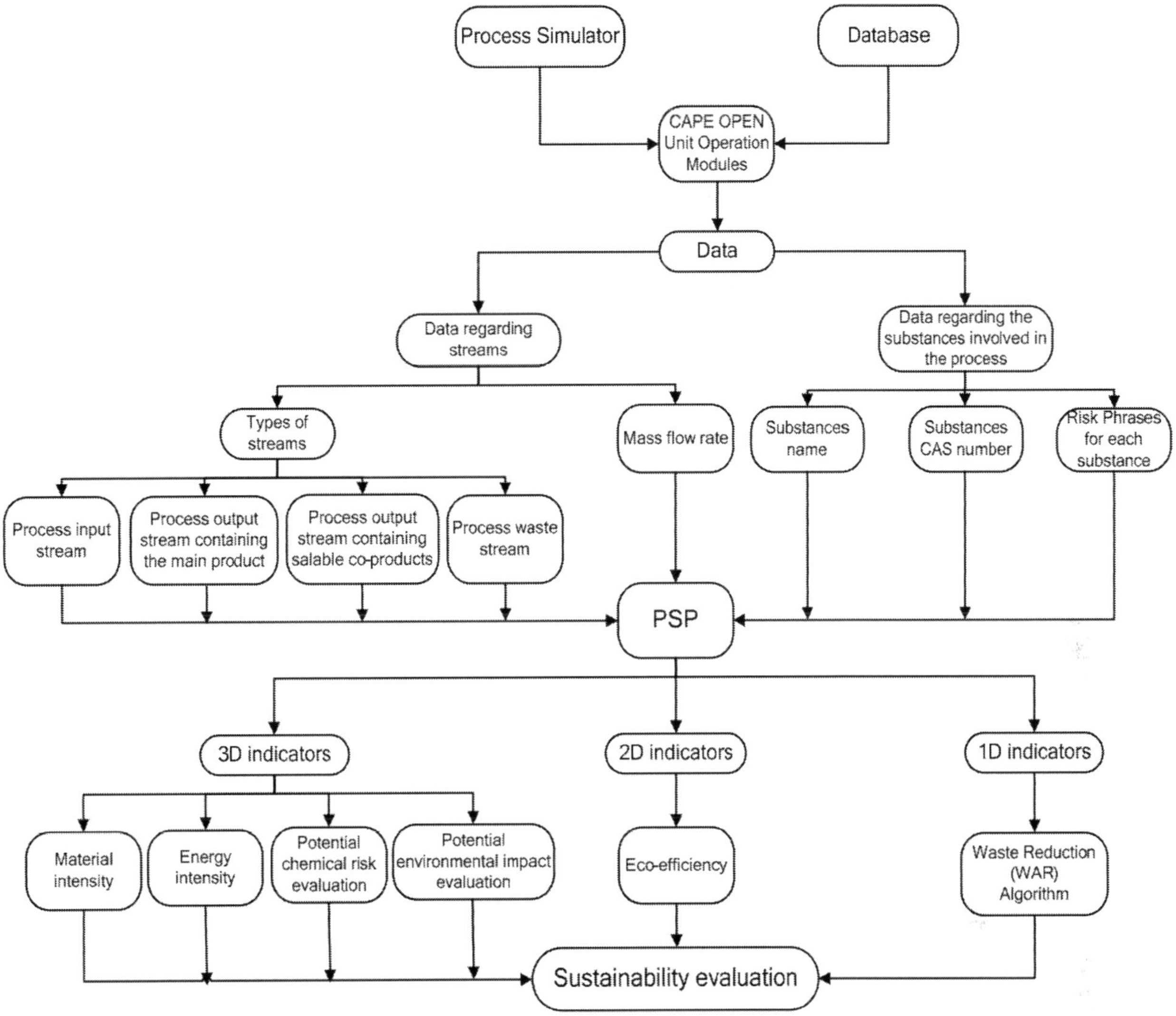


***Figure 1**: General Schema of PSP Framework for sustainability evaluation*

Details regarding 3D and 1D indicators and their implementation are given below.

As mentioned before, four different 3D indicators are considered. The MI represents the amount of non-renewable resources required to obtain a unit mass of products. The EI measures the process energy demands. The PCR considers risk associated with manipulation, storage and use of hazardous chemical substances in the process. The PEI is due to emissions and discharges of hazardous chemicals to the environment.

The implementation of the above mentioned indicators has been made according to Martins and Vincent (Martins *et al.*2006 and Vincent *et al.*, 2005). Some details regarding the two CO Modules for 3D indicators are given in the following paragraph.

As mentioned, two CO Modules have been developed to extract, from process simulator, the requested data for the 3D indicators calculation. The first module, called Promatrics, has four parameters, one input port and one output port. This module should be connected to the input and output process streams: its goal is to extract data for MI, EI and PEI calculation. The second CO Module, called Pro_Point_Risk, has only one parameter, one input port and one output port. This module should be inserted in a plant strategic point, the most dangerous or risky, in order to extract data for the potential chemical risk (PCR) evaluation.

The CO Modules' characteristics are summarized in Table 1. The parameter "name" is used to identify the CO Module. The parameter "connection" is a good indicator of the CO Module position in the flow-sheet. The "water_usage" and "air_usage" parameters are essential for the MI indicator calculation.

**Table 1**: Properties of the CO Modules used in 3D indicators

| CO MODULES NAME | PORTS | PARAMETERS | PARAMETERS VALUES | |
|---|---|---|---|---|
| Promatrics | One input port<br>One output port | name | From 1 to N, where N represents the number of the CO Modules (Promatrics.dll) inserted in the process flow-sheet | |
| | | connection | 0 | if ProMatrics is connected with a process input stream |
| | | | 1 | if ProMatrics is connected with a waste process stream |
| | | | 2 | if ProMatrics is connected with an output process stream |
| | | | 3 | if ProMatrics is connected with an output containing salable co-products |
| | | water_usage | 0 | if ProMatrics is connected with a stream in which water is not present |
| | | | 10 | if ProMatrics is connected with a stream in which water is present but is not included in the MI indicator |
| | | | 11 | if ProMatrics is connected with a stream in which water is present and is included in the MI indicator |
| | | air_usage | 0 | if ProMatrics is connected with a stream in which the air (oxygen and nitrogen) is not present |
| | | | 10 | if ProMatrics is connected with a stream in which the air (oxygen and nitrogen) is present but is not included in the material metrics |
| | | | 11 | if ProMatrics is connected with a stream in which the air (oxygen and nitrogen) is present and is included in the material metrics |
| Pro_Point_Risk | One input port<br>One output port | name | From 1 to N, where N represents the number of CO Modules (Pro_Point_Risk.dll) inserted in the process flow-sheet | |

After the process simulation, the data extracted by the CO Modules are processed in the developed software (PSP Framework) and the 3D indicators are calculated. A logical schema, containing all steps for the 3D indicators calculation, is presented in Figure 2.

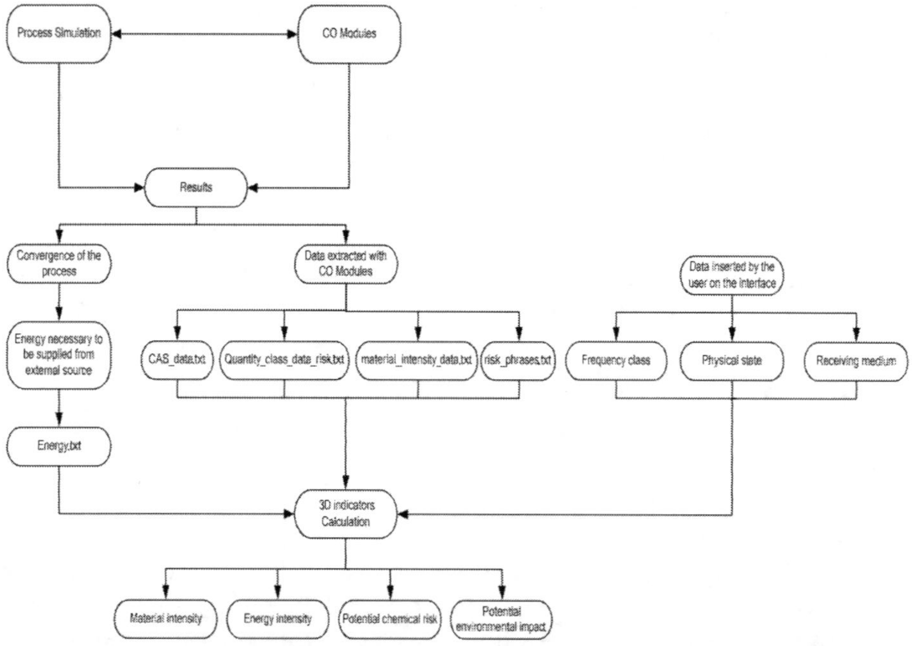


***Figure 2**: Logical schema for 3D indicators' calculation*

The access to the database (DB) is an important issue of the CO Modules. The DB contains risk phrases for more than 4500 substances.

The connection between the CO Modules and the database is made by using the ADO (ActiveX Data Objects) technology. The risk phrases search in the database is executed, as a function of the Chemical Abstract Service Number (CAS number), which is found in the process simulator. The access methodology is described in Figure 3.

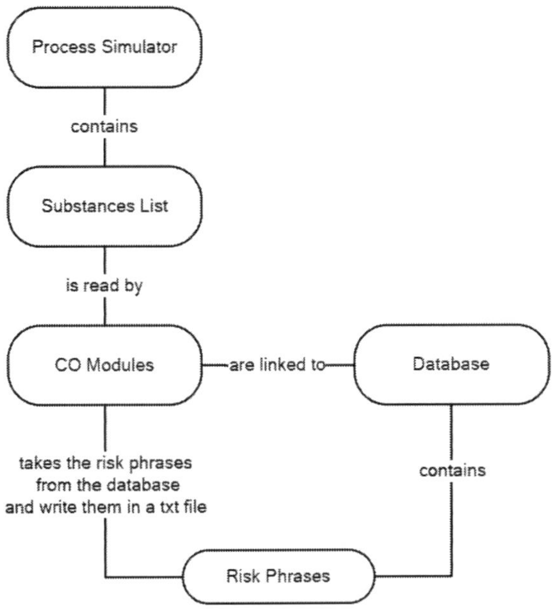


***Figure 3**: Logical Schema to access the database using the CO Modules*

Other two CO Modules have been developed for the 1D indicator. In the first module, called COWAR, the WAR Algorithm equations have been implemented. The COWAR has one input port, one output port and three parameters. The process type, position and name of the CO Unit Operation have been defined as real parameters. The values assumed by the parameters are presented in Table 2.

**Table 2**: Values assumed by the COWAR parameters

| CHEMICAL PROCESS | | ENERGY GENERATION PROCESS | |
|---|---|---|---|
| **Parameter Name** | **Parameter Value** | **Parameter Name** | **Parameter Value** |
| process_ type | 0 | Process type | 1 |
| cape_ position | 0 (for input stream) | CO position | 0 (for input stream) |
| | 1 (for waste stream) | | 1 (for waste stream) |
| | 2 (for output product stream) | | |
| | 3 (for internal stream of chemical process) | | |
| cape_ID | 0<br>1<br>.<br>N-number of CO modules inserted in the process | CO name | 0<br>1<br>.<br>N-number of CO modules inserted in the process |

With regard to the requirement of using three parameters, it should be specified that: the process type parameter is used to identify the process type (chemical or energy generation process); the position parameter is needed for the location detection of the CO in the flow sheet (the CO could be inserted on an input, output, intermediary or waste stream); and the name parameter is needed to find the CO unit operations, which have been first classified using the position parameter.

For a complete estimation of a chemical plant environmental impact, not only the chemical process, but also the energy process connected to the chemical process should be considered. The material and energy balance equations of coal combustion process have been implemented in a CO Module. This Module reads the necessary quantity of energy to be supplied from external sources. From the energy value, the Module calculates the gaseous emissions. The emissions stream is linked with the COWAR to calculate the impact of the energy process.

The PSP Framework was implemented in Visual Basic (version 6.0). The DB containing toxicological and risk phrases data was implemented in Microsoft Access. For the CO Modules development, the CAPE OPEN Unit Wizard (version 0.9.3, 1.0 and .NET) has been used.

## 3. Case Study: R-134a production process

The above presented methodology has been applied in several case studies in order to choose the most environmentally friendly process from different alternatives. The R-134a production process is reported as example.

The simplified schema of the process is presented in Figure 4.

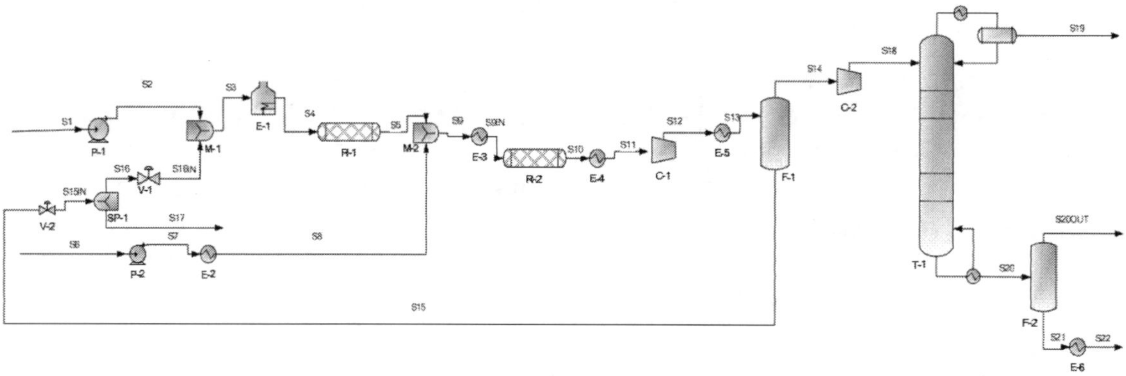


***Figure 4**: R-134a production process*

Process description:

Liquid hydrogen fluoride, S1, enters the system. It is pumped in P-1 before being mixed with a recycle stream (in M-1) consisting of HF, R-133a ($CF_3CH_2Cl$), R-134a ($CF_3CH_2F$), and trichloroethylene (TCE). The mixed stream, S3, enters in a heat exchanger E-1 where it is heated to 400°C before being fed to the first reactor R-1. The first reactor, R-1, is an isothermal plug-flow reactor which converts 99.3% of R-133a to R-134a. The product stream of the first reactor, S5, is then mixed with a feed of TCE, S8. The mixed stream, S9, is then fed through a heat exchanger, E-3, where it is cooled. The second reactor, R-2, converts HF and TCE to R-133a and hydrochloric acid (HCl). The stream leaving the second reactor, S10, is cooled in the heat exchanger, E-4, and sent to a compressor, C-1. The stream, S12, is then cooled to 30°C in E-5 and fed to the first flash, F-1. The distillate from this flash, S14, consists mainly of R-134a and HCl. This stream is after that compressed in C-2, before being fed into the distillation column T-2. Here, R-134a is removed from the rest of the refrigerants and the HCl. The 99.99% pure R-134a bottom stream, S20OUT, is cooled to 40°C before being storage (Turton *et al.*, 2003).

The process has been simulated using PROII process simulator.

Three design alternatives have been proposed and simulated for the process under study. The first case corresponds to the base case (Figure 4). In the second design, the reaction heat of the first reactor is removed from the process by means of a Dowtherm cooling loop. This design is depicted in Figure 5. In the third design, the heat integration between two heat exchangers (E-2 is a heater; E- 4 is a cooler) is considered. This design is illustrated in Figure 6.

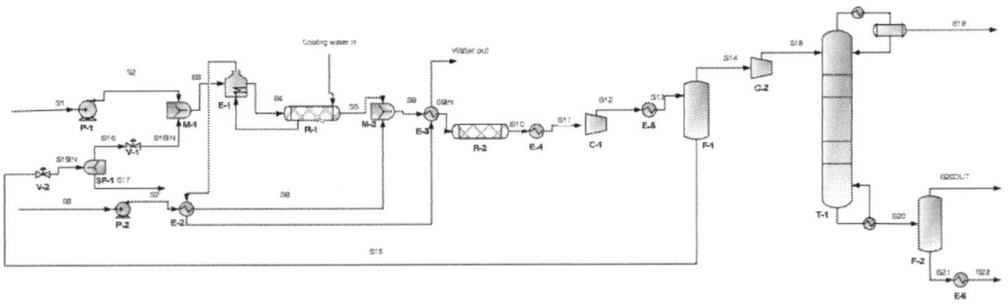

***Figure 5***: *R-134a production process (Case 2)*

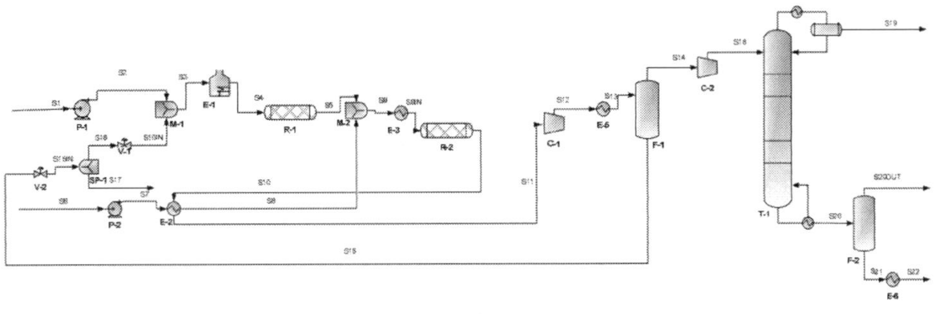

***Figure 6***: *R-134a production process (Case 3)*

The 3D indicators have been calculated using the data extracted with Promatrics and Pro_Point_Risk CO Modules. The results are reported in Table 3.

**Table 3**: 3D indicators obtained in the R-134a production process

| CASE STUDY | 3D INDICATORS | | | |
|---|---|---|---|---|
| | MI | EI | PCR | PEI |
| CASE 1 | 132.24 | 4.327 | 1300010 | 1889.3 |
| CASE 2 | 132.41 | 0.558 | 1300010 | 1889.3 |
| CASE 3 | 672.73 | 5.31 | 1300010 | 1889.3 |

Case number 2 has the lowest 3D indicators and therefore represents the most environmentally friendly design. For the same cases the WAR Algorithm (1D indicator) has been applied. The results are reported in Ta[illegible] 4. Case 3 is the most favorable. It can be noticed that the difference between case 2 and case 3 is not much, so the two cases can be considered equivalent. Making the comparison between the same case studies in Table 3, a big difference between these two alternative designs can be noticed. In conclusion, Case 2 is the most favorable design, taking into account 3D and 1D indicators. The results stress, once more, the necessity to use the complete PSP Framework.

**Table 4**: Environmental Impact Indexes for R-134a production process using WAR Algorithm

| CASE STUDY NAME | ENIVORONMENTAL INDEXES NAME | | | |
|---|---|---|---|---|
| | **Iout** (PEI/hr) | **Iout_mp** (PEI/kg) | **Igen** (PEI/hr) | **Igen_mp** (PEI/kg) |
| CASE 1 | 532.25 | 0.2566 | -10538.88 | -5.091 |
| CASE 2 | 489.99 | 0.2367 | -10580.14 | -5.111 |
| CASE 3 | 479.89 | 0.2318 | -10590.24 | -5.116 |

## 4. Conclusions

The present work describes the implementation of a framework used for sustainability evaluation. The software takes into consideration the aspects of sustainability during the design step of a chemical process. The framework background is represented by the 3D and 1D indicators found in the literature. The framework structure, the development of the CO Modules as well as the implementation are described in the present paper. The software has been tested in several cases and the R-134a production process is reported here.


## 5. Acknowledgment

The authors are especially grateful to: Branislav Opacic and Circe La Placa (ICS-UNIDO), Marco Carrone (University of Trieste), Ioan Toma (University of Innsbruck), who provided support and suggestions for the present work.


## 6. References


Cabezas, H., Bare, J.C., Mallick, S.K., 1997, Pollution Prevention with Chemical Process Simulators: The Generalized Waste Reduction (WAR) Algorithm, Computers and Chemical Engineering, 21, 305-310

Jenck, J.F., Agterberg, F., Droescher, M., 2004, Products and processes for a sustianable chemical industry: a review of achivements and prospects, Green Chem., 6, 544-556

Martins, A.A., Mata, T.M., Costa, C.A.V., Sikdar, S.K., 2006, Framework for Sustainability Metrics, American Chemical Society,

Turton, R., Bailie,R.C., Whiting,W.B.,Shaeiwitz,J.A., 2003, Analysis, Synthesis and Design of Chemical Processes, Upper Saddle River: Practice Hall, New Jersey, 946-957

Veleva V, Ellenbecker M., 2001, Indicators of sustainable production: framework and methodology. J Cleaner Prod; 9, 519–49

Vincent, R., Bonthoux, F., Mallet, G., Iparraguirre, J.F., Rio, S., 2005, Méthodologie d'Evalution Simplifiée du Risque Chimique: Un Outil d'Aide à la Décision. INRS Hyg. Secur. Travail, 195 (Second Quarter), 7 (INRS Ref. No. ND 2233)

# 398e Development of Cape-Open Unit Operations for Advanced Power Systems Modeling

**David A. Swensen**[1], Stephen E. Zitney[2], and Mike Bockelie[1]. (1) Reaction Engineering International, 77 West 200 South, Suite #210, Salt Lake City, UT 84101, (2) U.S. Department of Energy, National Energy Technology Laboratory, Morgantown, WV 26507-0880

Reaction Engineering International (REI) is developing a suite of CORBA CAPE-OPEN compliant unit operations to support the Department of Energy NETL Advanced Process Engineering Co-Simulation (APECS) framework for advanced power systems. One of these components will provide an interface to the GateCycle software, which is widely used in the power generation industry for process modeling. Additional CAPE-OPEN components will provide modeling capabilities of varying fidelity for equipment such as a gasifier, air separation unit, syngas cooler, solid oxide fuel cell and water-gas shift reactor. The models used in the components will be a combination of those developed by REI and proprietary models developed by members of the Project's industrial interest group.

# OPTIMIZER AS A CAPE-OPEN UNIT OPERATION

*Martin Gainville[1], Pascal Roux[2], Alain Vacher[3], Philippe Baudet[3], Michel Pons[4], and Didier Paen[5].*
*(1) Institut Français du Pétrole, IFP, 1-4 avenue de Bois Préau, RUEIL-MALMAISON, 92852, France, (2) Institut Français du Pétrole, IFP-Lyon, BP3, 69 390 Vernaison, France, (3) ProSim SA, Stratege Batiment A, BP 27210, Labege Cedex, 31672, France, (4) Michel Pons Technologie, 32 rue Raulin, Lyon, 69007, France, (5) RSI - Dynamic Simulation Solutions, Parc technologique de Pré Milliet, Montbonnot, 38330, France*

## Abstract

Design and operation of petroleum field facilities require more and more advanced flow calculations to optimize investments and maximize production and reserves. A number of integrated software platforms provide useful architectures to model multiphase flow from well-bore to topside facilities. Even through the use of such modeling technologies, the complexity of oil & gas production systems remains a strong obstacle to manually screen the system settings for optimal operation and design: the use of an optimizer is well suited in that case. The paper discusses the integration of Sequential Quadratic Programming optimization module developed by ProSim SA (thereafter called OPTI) into the CAPE-OPEN compliant simulation environment INDISS™ developed by RSI. OPTI resolves non-linear optimization problems. For its integration into INDISS™, OPTI has been wrapped as a CAPE-OPEN unit operation (called CO-OPTI) that can be linked with other unit operations on the flowsheet thanks to CAPE-OPEN (CO) standard interfaces. An example based on a deepwater production system illustrates the benefit of this development.

## Introduction

The TINA project[1] is a R&D project set up by TOTAL and IFP to prove the feasibility of integrating complex physical models, coming from different software providers, into a simulation platform while simultaneously insuring data and model consistency. Within the TINA project, the integration of an optimization module into the INDISS™ process modeling environment[2] has been investigated. As the INDISS™ platform is compliant with CAPE-OPEN Thermodynamic and Physical Properties interface specification as well as with CAPE-OPEN Unit Operations interface specification, a pragmatic way for integrating an optimizer is to wrap it as a CAPE-OPEN unit operation. Through CAPE-OPEN standard interfaces,, the module can then be linked with other unit operations in order to solve non linear optimization problems on complex process systems. The paper presents how the OPTI[3] optimizer has been wrapped as CO-OPTI and then integrated into INDISS™. Previous work[4] was aimed at wrapping the SPEC module developed by ProSim SA. SPEC resolves design specification problems by a sequential modular approach. Wrapped as a CAPE-OPEN Unit Operation, SPEC was easily used within INDISS™. Work on SPEC proved that it was feasible to make such a module a CAPE-OPEN compliant Unit Operation with the objective of easily integrated it within INDISS™. It paved the way to the work on CO-OPTI.

An application example is provided to show that a dedicated CAPE-OPEN unit operation is able to optimize an objective function under constraints. The use of the CO-OPTI component is not reserved only to the oil & gas upstream domain that is the context of the TINA project. CO-OPTI has been shown to work exactly the same way in two different process modeling environments (PMEs) applicable to other process industries, and CO-OPTI may be of interest to a number of organizations for easily integrating optimizers into CAPE-OPEN PMEs.

## Optimization approach

A sequential modular approach is used in INDISS$^{TM}$ to resolve recycle variables under steady state conditions. A production system is modeled as a network of units linked by material streams and information streams. The optimization problem is made of the network equations expressed in terms of equality and inequality constraints (including those relative to tear streams), and involves also the minimization of an associated objective function .

Two different approaches are usually used to define and solve such optimization problems: either the network equations are solved separately at each iteration of the optimizer (this method is called "feasible path"), or the network equations, inequality constraints and objective function are combined and solved within the same iterative stage (this method is called "unfeasible path"). The "unfeasible path" approach has been chosen in the present study as it is robust and less time consuming and also as it makes easier the implementation of an external solver into an existing platform. The main issue of the OPTI integration into INDISS$^{TM}$ is to define the right protocols for ordering calculation and transferring information from the process model to the optimizer and back.

## Optimizer description and integration

The OPTI module is an internal solver of the ProSimPlus platform[3] developed by ProSim SA and its Non Linear Programming algorithm is based on work done at Carnegie Mellon University.[6] The OPTI module is aimed at solving non linear optimization problems and has been specifically adapted to solve flowsheet problems using a modular sequential approach. The OPTI updates iterative values composed by tearing streams (temperature, pressure and partial flow rates) and by operating and design parameters of the modules (for example valve openings, pipe diameters, pump rotation speeds). Then, sequential calculations are performed to evaluate equality constraints, inequality constraints, the objective function and their respective gradients. Subsequently a new set of iterative values are proposed for the next iteration. Thus, the process model is seen as a black-box by the optimizer. One issue with this approach is the number of sequential calculations required to estimate derivatives by numerical differentiation.

The INDISS$^{TM}$ platform is compliant with CAPE-OPEN thermodynamic and unit operation interface specifications.[2] Unit operations refer here to process units such as pipes, valves, separators and pumps. Methods of the CAPE-OPEN Unit Operation interface specification allow initializing unit operations, connecting them to each other by material streams and information streams. Thus, it appears that CAPE-OPEN Unit Operation interfaces may be implemented on CO-OPTI in order to plug OPTI within INDISS$^{TM}$ and then to run optimization calculations. The integration of CO-OPTI into INDISS$^{TM}$ requires data transfer back and forth between the optimizer and the rest of the flowsheet. A number of CAPE-OPEN information streams are used to send back to CO-OPTI how the equality and inequality constraints are satisfied and to get back the value of the objective function (Figure 1). Other CAPE-OPEN information streams update iterative variables within the flowsheet components. In addition, tear material streams are connected to CO-OPTI. This allows CO-OPTI to manage the convergence of tearing problems. A sequential calculation graph is established by INDISS$^{TM}$ based on the flowsheet topology and by taking into account the role of CO-OPTI. Usually,

CO-OPTI is placed in the last position of the calculation sequence when all process equipment modules have proceed with their internal calculation and have been updated. In addition, CO-OPTI communicates with INDISS™ using a specifically named CAPE-OPEN parameter that represents a convergence flag which has to be checked by the PME in order to stop calculations when convergence is achieved or not. Some specific developments have been implemented within INDISS™ to order the sequential calculation list and to deal with the CO-OPTI specific parameter. The Barkeley and Motard algorithm[7] has been used to ensure a proper calculation order with respect to partitioning and tearing. For the configuration given in Figure 1, the calculation order is {UO1, UO2, UO3, Evaluation module, UO5, UO4, CO-OPTI}.

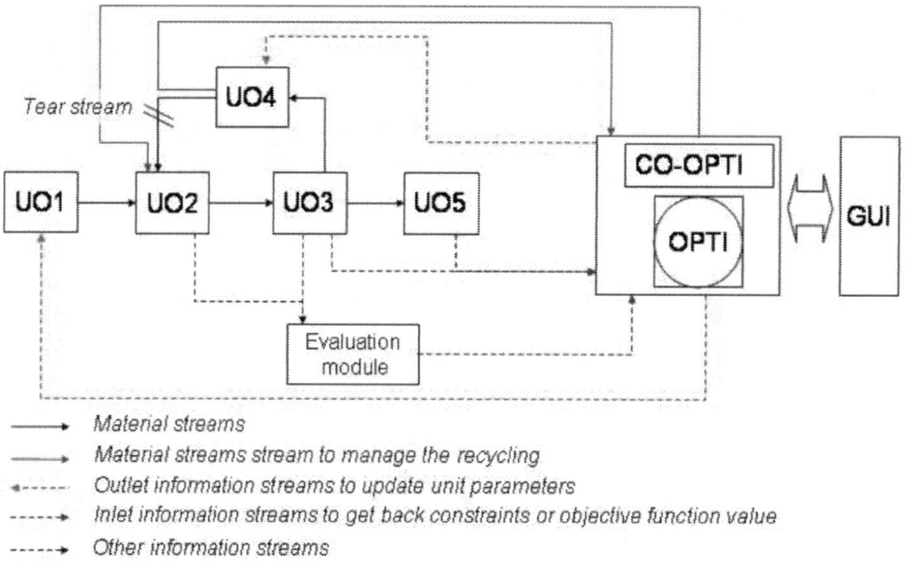


**Figure 1:** Schematic view of the CO-OPTI optimizer and the CAPE-OPEN process modeling environment

ProSim SA has a strong experience in developing CAPE-OPEN software components.[8-9] Thus, ProSim SA has been selected to make OPTI CAPE-OPEN compliant. For validating developments, cross-comparison tests have been conducted on three integration configurations: OPTI within ProSimPlus, CO-OPTI within ProSimPlus using CAPE-OPEN unit socket developed in ProSimPlus and then CO-OPTI with INDISS™. On a linear optimisation case based on simple process model (Figure 2), OPTI and CO-OPTI have provided exactly the same results at each numerical iteration for the three configurations.

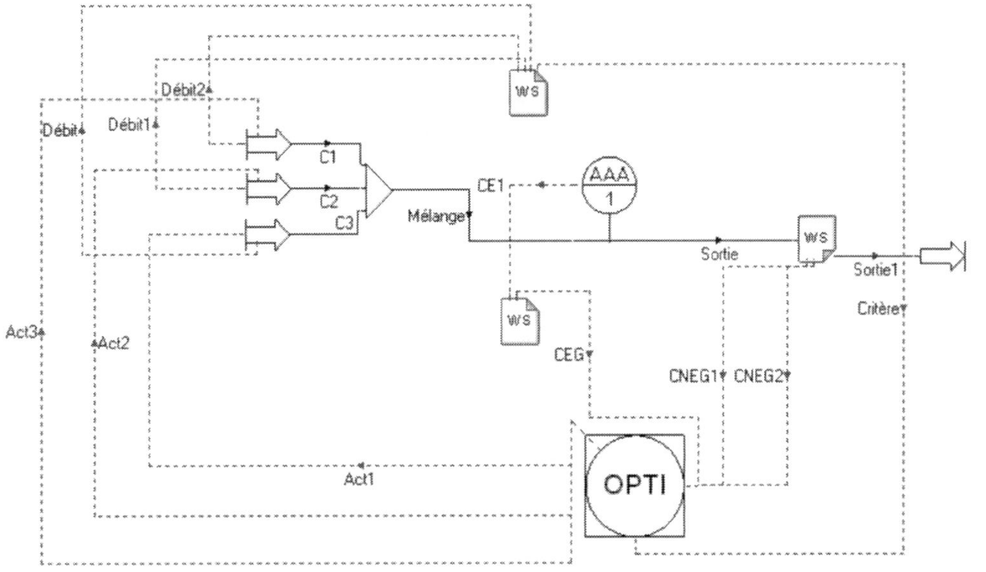


**Figure 2:** Simple process model for validating CO-OPTI integration on a linear optimization problem (ProSimPlus screenshot)

## INDISS™ Process Modeling Environment

The INDISS™ platform allows defining mixed networks composed of both native and third-party CAPE-OPEN unit operations. In the TINA project, CAPE-OPEN unit operations have been developed to calculate steady state flow in pipes. Hydrodynamic models such as TACITE hydrodynamics[10], are used to solve accurately pressure losses and phase velocities within pipes. In addition, INDISS™ implements a CAPE-OPEN thermodynamics socket to plug CAPE-OPEN Physical Property Package. The set of unit operations available within INDISS™ and the choice of an appropriate Physical Property Package allow modeling of oil & gas production systems under steady state conditions with a fully compositional approach. Moreover, INDISS™ is aimed to solve transient problems defined on the same flowsheet by using a specific solver and dedicated CAPE-OPEN dynamic unit interfaces[11].

## Case study

A case study has been defined to show how CO-OPTI can be used within INDISS™ and which useful information can be obtained by performing optimization on a deepwater production system using the software components described previously.

***System layout***

The sub sea system is an oil dominated development representative of a West African deep offshore field. The system is composed of two sub-sea wellhead clusters respectively with two and three wells (Figure 3). Fluid flows out of wells are controlled by choking wellhead valves. Clusters are connected together by a sub sea flow line and a second flow line, connected to a riser, transports the production up to surface facilities. A sub sea multiphase pump unit is installed downstream the second manifold to boost the production. The second sub-sea line and the riser have the same inner diameter that has been initially set to eleven inches pipes. Finally, a simplified surface process is considered to separate oil and gas phases. The average water depth is 1500 meters high.

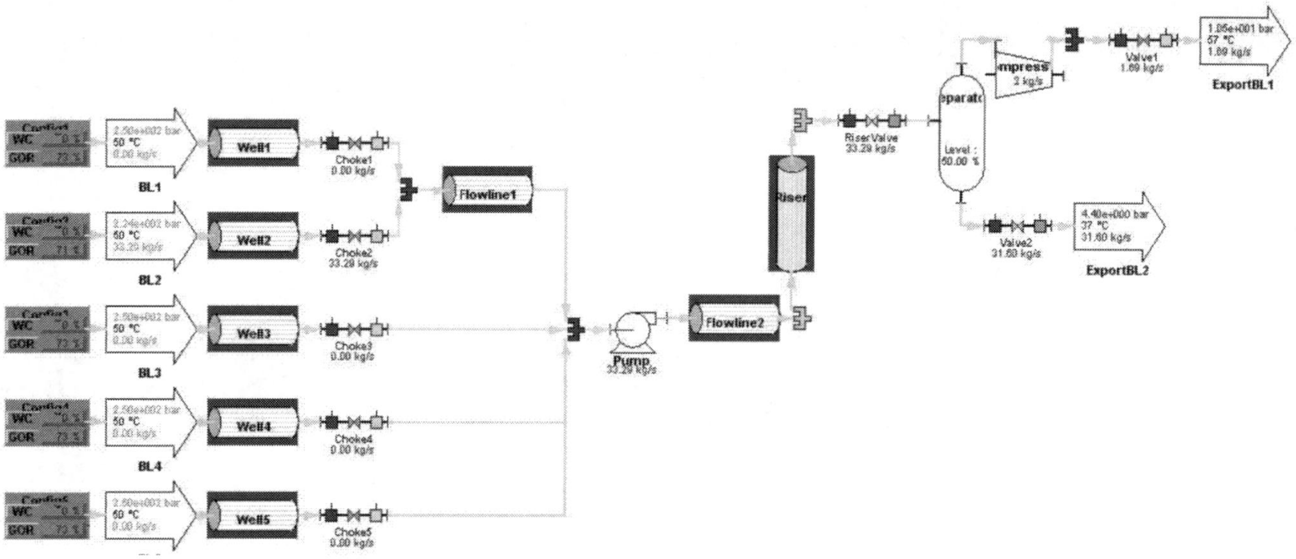


**Figure 3:** System layout with the two sub sea clusters, the multiphase pump and the basic surface process (INDISS™ screenshot)

***Optimum design cost for several tie-back distances***

Optimization runs have been performed on the sub sea system with the objective of minimizing the design cost (CAPEX). The production system should be designed to allow production over the field lifecycle and the most restrictive period is often used for sizing the sub sea pump unit. Usually, this period corresponds to the end of the production of the field when a large amount of water is produced due to water breakthrough that consequently increases hydrostatic pressure losses. So, the most restrictive period has been chosen to set well input conditions (temperature, pressure, composition, gas-oil ratio, watercut, standard oil production rate) and to perform the design optimization.

The objective function to minimize is the overall cost which is defined by the sum of the pipe installation cost and the sub sea pump unit cost. The pipe installation cost is mainly depending on the pipe length and the pipe diameter and the pump cost is directly correlated to useful power to boost the production. For the present study, cost functions for the line and the pump are given by Equation 1 and Equation 2 and the pump useful power is given by Equation 3.

$$C_{Line} = L\left(\alpha D^2 + \beta D + \gamma\right) \quad (1)$$

$$C_{Pump} = A + B W_{Pump}^{C} \quad (2)$$

$$W_{Pump} = Q \ \Delta P \quad (3)$$

Optimizations have been performed using CO-OPTI with the following iterative variables: the diameter of the sub sea line from the pump unit to the riser base and the useful power of the pump. The operating constraint is an arrival topside pressure equal to 15 bars. Inequality constraints correspond to a maximum useful pump power of 1.2 megawatts (equivalent to a maximum differential pressure of 93 bars) and a pipe diameter range from 8 inches to 12 inches. Optimizations have been performed using the INDISSThermo Thermodynamic Package[2]. The tie-back distance has been varied from 5 km to 25 km to study the dependency of the optimum cost with the flow line length. To design line diameter and pump unit capacity, three options have been considered:

- Option 1: system designed with the maximum specified pipe diameter (Pipe diameter of 12 inches),
- Option 2: system designed with the maximum pump unit capacity (Pump unit of 1.2 MW),
- Option 3: the optimum solution calculated using CO-OPTI module (Optimum solution).

For options 1 and 2, one of the iterative variables is fixed at its upper bound and the other is calculated to satisfy the operating constraint of 15 bars at the arrival. Calculations have been made using SPEC module[3]. This provides (diameter, power) solution vectors that satisfy production constraints but that are not cost optimal. Hereafter, the solutions are compared for several tie-back distances.

The overall cost increases with the tie-back distance (Figure 4) due to the higher pressure required to boost the production and also to the higher pipe length to be installed. For distances up to 25 km, the optimum cost solution calculated by CO-OPTI (option 3) gives a useful pump power at the upper bound (1.2 megawatts): this corresponds to install a pump of maximum capacity and the solution is identical to option 2. In Figure 5, the cost reduction in percentage of the overall optimum cost is presented. For small distance (5 km), an over-sized pump unit is not suitable and an optimum solution (power unit of 0.61 megawatts and pipe diameter of 9.92 inches) allows reducing the overall cost by 10.2 per cent. Also, an optimum pipe diameter instead of the 12 inches diameter allows reducing the overall cost by 2.1 per cent. Then, for higher transportation distance, a high power pump unit becomes necessary and a cost benefit may only be expected on pipe diameter. Figure 6 provides the cost distribution between line installation and pump unit for the optimum case (option 3). Figure 7 and Figure 8 show evolutions of the pump useful power and the line diameter for the three options.

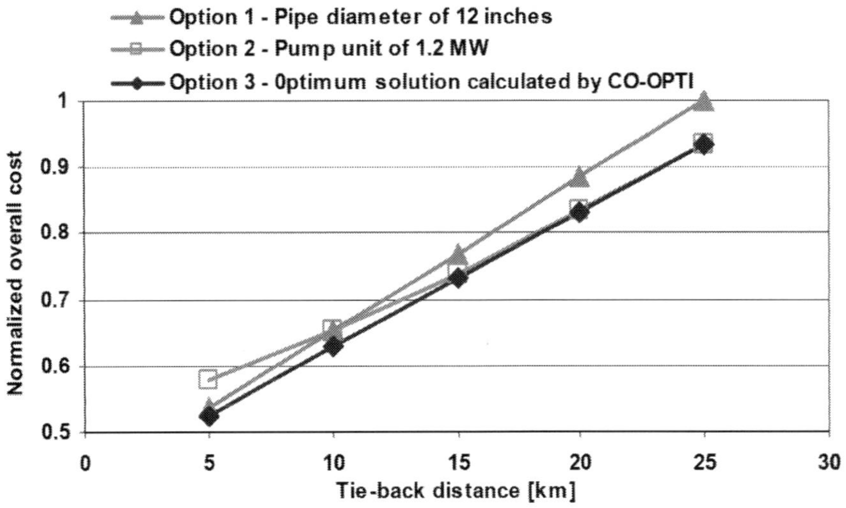


**Figure 4:** Overall cost evolution for the three design options

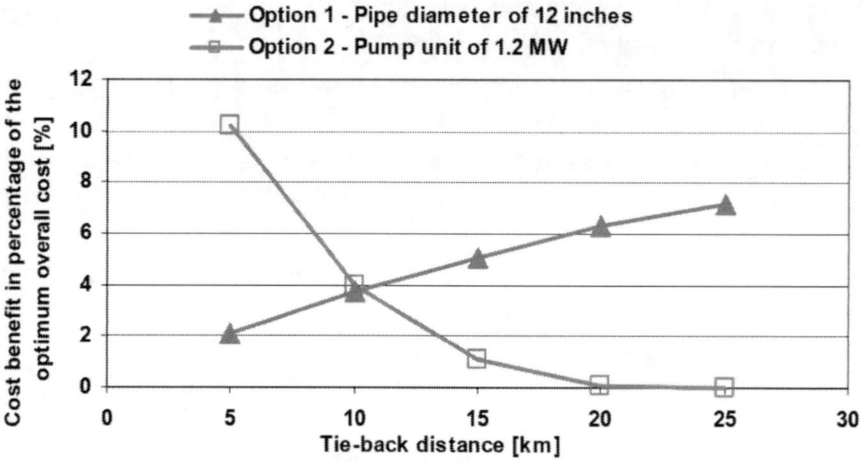


**Figure 5:** Cost benefit in percentage of the optimum overall cost for the 1.2 MW pump option and the 12 inches diameter option

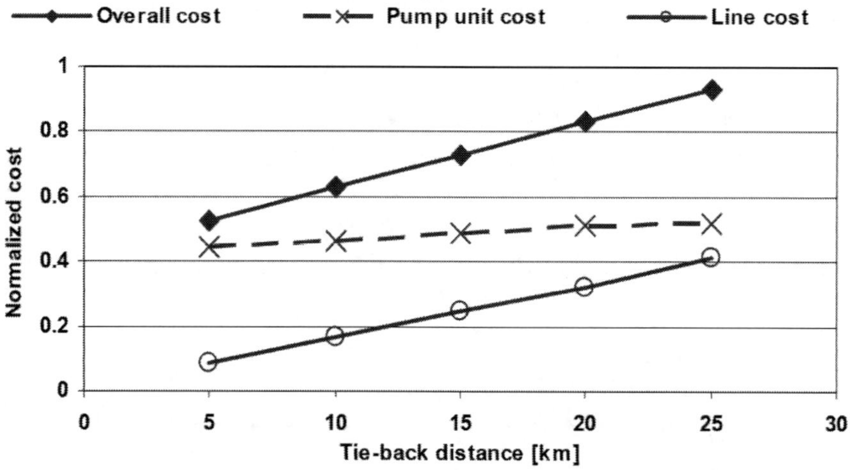


**Figure 6:** Cost distribution between the pipeline installation and the pump unit for the optimum case

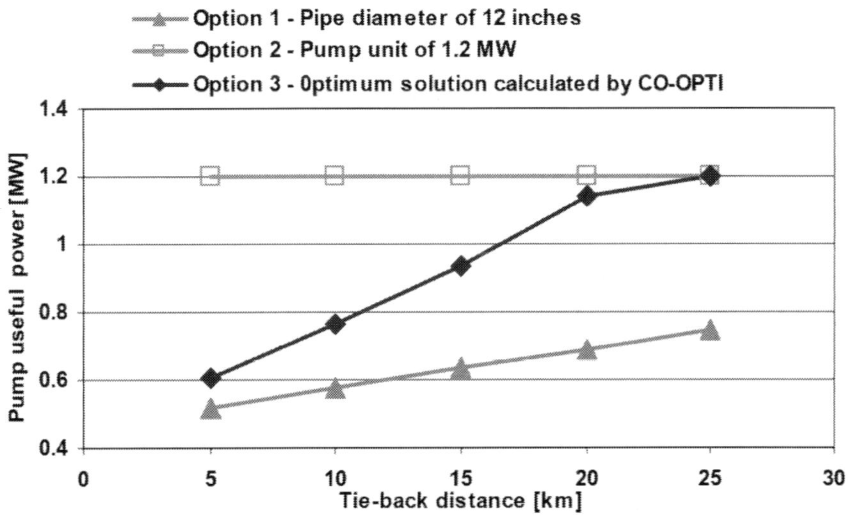


**Figure 7:** Pump useful power for the three design options

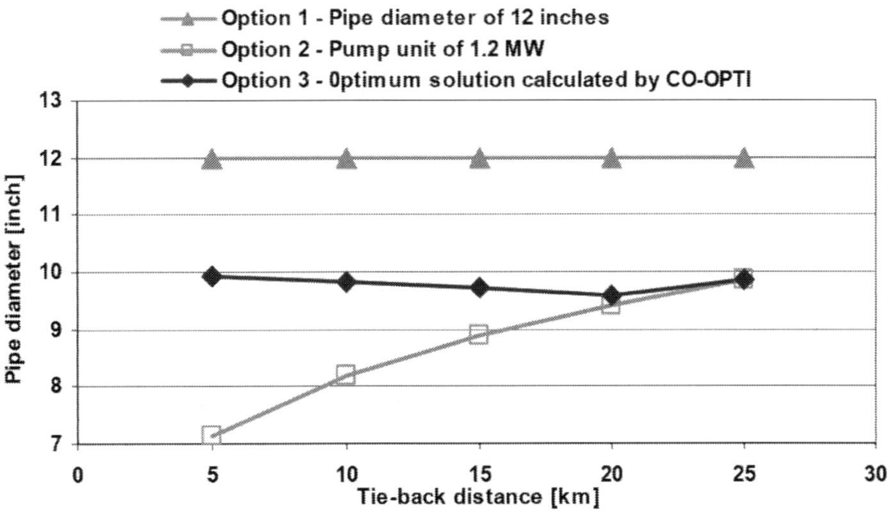


**Figure 8:** Pipe diameter for the three design options

The optimization results are strongly dependent on the system layout, boundary conditions and cost function definition. For example, changing production conditions will provide new information on the system which will lead to review the design solution. Optimization studies should be led with care even though integrated software platforms propose more and more user-friendly workflows to assist users in their works. In addition, for designing production systems, it is often necessary to take into account the production constraints over the production life, especially the increasing water cut, and multi-periods optimization including operating costs has to be setup. This has been studied in the TINA project but it is still a challenge for further developments.

## Conclusion and perspectives

Using CAPE-OPEN technology, it has been possible to integrate with success a third party optimizer as a unit operation in a compliant PME without huge and complex developments. An optimization study has been conducted using the interoperability arrangement set up and has provided some interesting results on a deep water oil production system. This solution may be of interest to a number of organizations. For further developments, IFP has planned to integrate its in-house optimizer within INDISS™ by using the same integration technique.

## Acknowledgements

We are grateful to TOTAL for their contribution and support to this work, RSI for having made available the INDISS™ platform and ProSim SA for the work done on OPTI module.

## Nomenclature

| | |
|---|---|
| $L$ | *Pipe length [m]* |
| $D$ | *Pipe diameter [m]* |
| $C_{Line}$ | *Line installation cost [MUSD]* |
| $C_{Pump}$ | *Subsea pump unit cost [MUSD]* |
| $\alpha, \beta, \gamma$ | *Cost coefficients relative to line installation* |
| $A, B, C$ | *Cost coefficients relative to pump installation* |
| $W_{Pump}$ | *Pump useful power [W]* |
| $Q$ | *Volumetric flow rate [m3/s]* |
| $\Delta P$ | *Pump deferential power [Pa]* |

## References

1. Gainville, M., Roux, P., M. Pons., M., Ricordeau, A., Paen, D., "Integrated Flow Modeling Platform Using CAPE-OPEN Standard RSI - Dynamic Simulation solutions", paper to be presented at the SPE Annual Technical Conference and Exhibition, Anaheim, USA (2007).
2. RSI, INDISS™ platform and INDISSThermo Thermodynamic Package, RSI Dynamic Simulation Solutions, Montbonnot, France (2007).
3. ProSim, ProSimPlus Process Modeling Environment, Simulis® Thermodynamics, OPTI optimizer, SPEC design specification solver, ProSim SA, Labège, France (2007).
4. Floquet, P., Joulia X., Vacher A., Gainville M., Pons M.: "Pressure-driven Steady-State Simulation of Oilfield Infrastructure", paper presented at the 17th European Symposium on Computer Aided Process Engineering (ESCAPE-17), Bucharest, Romania, V (2007), V.Plesu and P.S. Agachi (Eds), Elsevier B..
5. CAPE-OPEN Open Interface Specifications: "Thermodynamic and Physical Properties", "Unit Operation", available on CO-LaN website (2002).
6. Biegler, L.T., Cuthrell, J.E.: "Improved infeasible path optimization for sequential modular simulators – II: the optimization algorithm", Comput. Chem. Engng (1985) 9.
7. Biegler, L.T., Grossmann, I.E. and Westerberg A.W.:"Systematic Methods of Chemical Process Design", Prentice-Hall Inc., NJ (1999).

8. Vacher, A., Dechelotte, S., Baudouin, O.: "CAPE-OPEN Unit Operations: Development and Usage", paper presented at the 3rd US CAPE-OPEN Conference, topical conference of 06'AIChE, San Francisco, USA (2006).
9. Morales-Rodriguez, R., Sales-Cruz, M., Gani, R., Dechelotte, S., Vacher, A., Baudouin, O. : "Interoperability between Modelling Tools (MoT) and Process Simulators (ProSimPlus) through CAPE-OPEN Standards", paper presented at the 3rd US CAPE-OPEN Conference, topical conference of 06'AIChE, San Francisco, USA (2006).
10. Pauchon, C., Dhulesia, H., Lopez, D. and Fabre, J.: "TACITE: a comprehensive mechanistic model for two-phase flow", paper presented at the 6th BHRG Multiphase International Conference, Cannes, France (1993).
11. Paen, D., Cameron, D., Colantonio, C.: "Enhancing Process Modelling Environments: CAPE-OPEN Dynamic Unit Operations", Poster C44-001 presented at the 7th World Congress of Chemical Engineering, Glasgow, UK (2005).

# 398g Importing and Using Com-Based Co Compliant Unit Operations in Usepa Metal Finishing Pollution Prevention Tool

William M. Barrett Jr., National Risk Management Research Laboratory, US Environamental Protection Agency, Cincinnati, OH 45268 and **Svetlana Strunjaš-Yoshikawa**, Computer Science, University of Cincinnati, Cincinnati, OH 45221.

In this presentation interoperability issues between .NET-based US Environmental Protection Agency's Metal Finishing Pollution Prevention Tool (USEPA MFP2T) and CAPE-OPEN compliant, COM-based unit operations will be discussed. USEPA MFP2T is a sequential modular process simulator, built upon CAPE-OPEN interfaces using .NET framework. The CAPE-OPEN interfaces that are used in the simulator are imported from CAPE-OPEN COM type libraries. USEPA MFP2T is created using Microsoft Visual Studio C++ 2005.

In order for USEPA MFP2T to import and use CAPE-OPEN compliant COM-based unit operations, it is of a crucial importance to enable interaction and interoperation between this .NET-based application and COM-based components. The most common way for using COM objects from .NET applications is through a runtime callable wrapper (RCW), which enables a .NET client to access to a COM server. The RCW wraps COM objects and mediates between them and .NET common language runtime (CLR) environment, making the COM objects appear to .NET clients just as if they were native .NET objects, and making the .NET clients appear to the COM objects just as if they were standard COM clients. USEPA MFP2T contains .NET wrappers that enable import and use of COM-based, CAPE-OPEN compliant unit operations inside and these wrappers will be explained in details in this presentation.

Functionality of the .NET wrappers mentioned above is successfully tested by importing and using some COM-based unit operations from COCO simulator and results will be shown in this presentation.

## TD 87550: Cape-Open Thermo 1.1 Interface Support for Aspen HYSYS

Suphat Watanasiri, Aspen Technology, Inc., Ten Canal Park, Cambridge, MA 02141-2201

In this paper, we will describe the implementation of the CAPE-OPEN Thermodynamic and Physical Properties Interface version 1.1 within the process simulator, Aspen HYSYS, version 2006.5. The implementation includes both the socket and the plug so that Aspen HYSYS can use any v1.1-compliant thermodynamic package in its simulation calculations (socket). It can also export a v1.1-compliant thermodynamic package for use with any 3rd party Process Modeling Environment that supports this version of the thermodynamic interface (plug). As with v1.0 of the CAPE-OPEN plug, the v1.1 plug uses Aspen COMThermo as the physical property engine. Aspen HYSYS also continues to support the CAPE-OPEN Thermo 1.0 interface. An independent test harness was also created to test all the interfaces implemented by the plug and any v1.1 thermodynamic packages.

Interoperability tests with other v1.1-compliant thermodynamic packages will be discussed.

# 459b Experiences in Interfacing the Ik-Cape Thermodynamics Package to Cape-Open

**Lars Von Wedel**, AixCAPE, Intzestrasse 1, Aachen, Germany

In the late eighties, a consortium of german companies called IK-CAPE developed a specification of routines to calculate thermophysical properties in FORTRAN. An implementation according to this specification was also developed and used in several in-house simulation codes and academic research projects. In a joint project with an industrial partner, AixCAPE has equipped this IK-CAPE package with a CAPE-OPEN thermodynamic interface to permit the use of the existing property data input files in state-of-the-art simulation environments.

The presentation explains the capabilities and legacy interfaces of the IK-CAPE package and describes technical aspects of the migration process. Further, experiences made with the CAPE-OPEN-compliant package obtained from this migration effort will be shared. Successful tests have been made with AspenPlus, Pro/II, COCO, gPROMS, and Simulis and will be reported.

# 459c Development Of Tuwax As A Cape-Open Compliant Process Modelling Environment

Alain Ricordeau[1], Cem Sarica[2], and **Anais Mathieu**[1]. (1) TOTAL, PAU, 64000, France, (2) Petroleum Engineering, University of Tulsa, TULSA, OK 74110

The primary objective of this project, founded by TOTAL, is to modify TUWAX source code to comply with the CAPE-OPEN (CO) standards and so to enable TUWAX as a standalone application capable to run with any commercial thermodynamic software using 1.1 CAPE-OPEN Thermodynamic & Physical Properties interfaces.

1) Presentation of TUWAX

• TUWAX is a program for predicting paraffin deposition in single and two-phase flow environments. It is developed at Tulsa University Paraffin Deposition Project (TUPDP) and is regularly updated. Indeed knowledge gained from experimental modeling studies is utilized to permanently enhance the computer program written in FORTRAN 77.

• The wax deposition computer simulator consists of four main elements:

- The hydrodynamic module is used to calculate pressure gradient and to identify the flow pattern existing in a section of a pipeline.

- Depending on the flow pattern predicted previously, the heat transfer module predicts the bulk and wall temperature along the pipe segment.

- The thermodynamic module, developed by Multiphase Solution Inc. (MSI), predicts the wax concentration and properties of existing phases at specific temperatures and pressures.

- The wax deposition module predicts the wax deposition rates from the prevailing flow and temperature conditions.

• The minimum input data are:

- Pipe configuration (Diameter, Length, Pipe Conductivity, Surrounding Temperature, etc.)

- Operating conditions (Inlet Temperature and Pressure, Flow rate)

- Fluid properties (currently obtained from MSI-TU X)

• Relationship between TUWAX and MSI's thermodynamic package:

MSI generates some look-up tables of thermodynamic properties that are required for the Wax deposition calculation. The user must input the fluid composition, the pressure range and number of intervals; the temperature range and the number of intervals; and the model to use. Then, TUWAX is able to read and interpolate these tables by using appropriated routines.

2) Development of TUWAX standalone application

• The objective is to plug any commercially available CAPE OPEN Compliant Thermodynamic Software into TUWAX.

• To reach this objective, we used a library (COThermoWrapper1.1.dll) developed by PROSIM SA.This library allows an easy access to CAPE-OPEN Thermo Property Packages 1.1 from an application written in FORTRAN. This application can be a standalone program or a CAPE-OPEN Unit Operation. The library acts as a wrapper around the TUWAX code adding support of CO interfaces.

• The wrapping approach preserved the main three modules of the code (Hydrodynamic, Heat Transfer, and Wax Deposition). We only modified the routines in charge of reading and interpolating the look-up tables, which represents 15 % of the code. Thus, it will be easy to update the other three modules described previously if needed.

3) Interoperability with TUWAX using Infochem's MultiFlash thermodynamic services

• Some tests using an Infochem's Multiflash Property Package have already been conducted on an industrial case provided by TOTAL and allowed us to verify the compatibility between TUWAX and a CO compliant Thermodynamic Software through 1.1 Thermo Interface.

• Simulations with MultiFlash are intended to ensure a better fluid description because we can keep track of the change in composition in time and space. The water phase will also be better described and that may give us an opportunity to enhance the TUWAX model.

## Simulating Wax Deposition in Pipelines with a CAPE-OPEN 1.1 Thermodynamics Package

*Beryl Edmonds, Tony Moorwood, Richard Szczepanski* and Xiaohong Zhang*

*Infochem Computer Services Ltd., 13 Swan Court, 9 Tanner Street, London SE1 3LE, UK.*

** Corresponding author. rs@infochemuk.com*

### Introduction

Wax deposition from petroleum fluids is a common problem in the oil industry. It affects principally the production and transportation of crudes but may also occur with lighter fluids such as condensates. In this paper we outline the modelling components necessary to simulate wax deposition in a pipeline. Particular attention is focused on the physical property requirements for such a simulation. The situations considered range from the simplest case of an oil pipeline to the increasingly common situation where oil, gas and water phases are present in addition to the wax.

As shown in Figure 1, successful simulation of wax deposition requires three principal software components:

1. a model for multi-phase fluid flow in the pipeline which need only use quasi-steady-state techniques;
2. a model for the rate of wax deposition on the pipeline walls;
3. a thermodynamic and physical property model package for the wax and fluid phases.

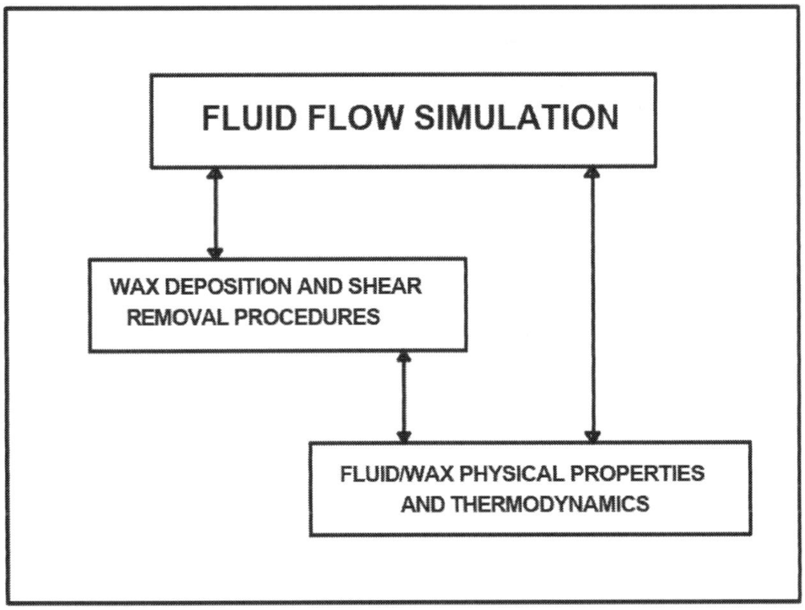


Figure 1. Components of a wax deposition simulator.

A new simulator, FloWax, has been developed that covers (1) and (2) which in turn makes use of the Multiflash thermodynamic and physical properties package. We have now added a CAPE-OPEN physical properties interface so that any CO 1.1 compliant property package can be used.

## Modelling principles

During the development and validation of the FloWax package, it was found that there are four important aspects of the deposition model that need to be made physically realistic in order to obtain good agreement with flow loop experiments and also give predictions for full-scale pipelines that are in accord with operating experience. These are the thermodynamic model for wax precipitation, the need to use a fully compositional description of the wax phase, the method of calculating the wax deposition rate and the shearing effect of turbulent flow in reducing or preventing wax deposition.

## Thermodynamics

The thermodynamics of the wax phase is important as it defines the thermodynamic driving force which controls the rate of wax deposition. The thermodynamic model should not only define the wax appearance temperature (WAT) accurately, but needs also to give an accurate description of the whole wax precipitation curve. Figure 2 gives an example for an oil cited by Erickson et al [1]. As the figure illustrates, the predictions given by different thermodynamic models vary considerably. The model we prefer is the Coutinho model which is implemented in the Multiflash program, and has been discussed by Coutinho et al. [2].

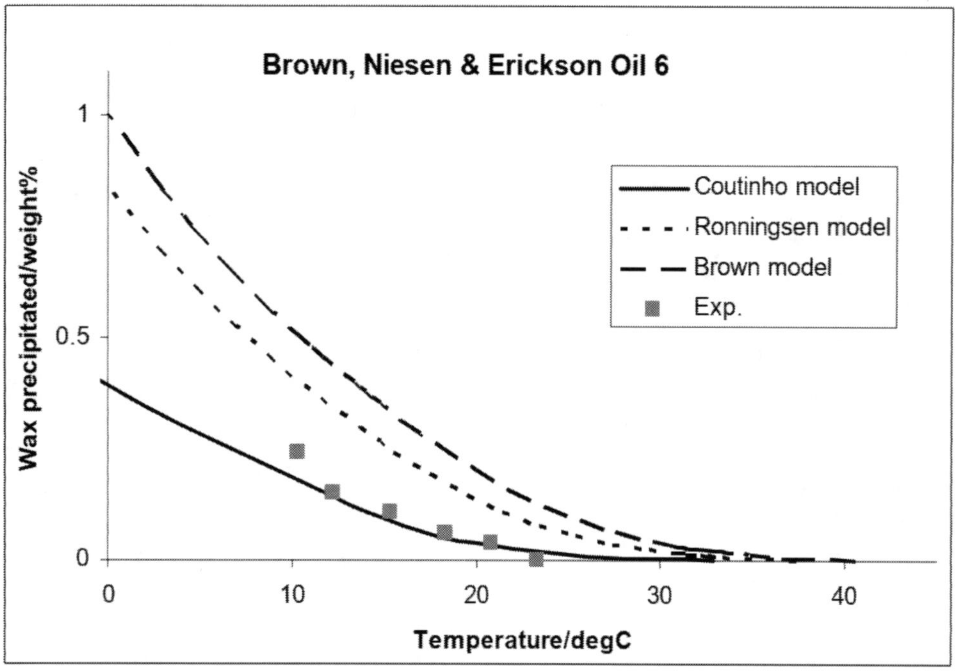


Figure 2. Wax precipitation curve for Brown et al. Oil 6

## Compositional treatment of wax phase

Waxes are crystals of n-paraffins which occur as a discrete distribution up to high molecular weights. A rigorous simulation of wax deposition therefore typically requires something approaching 100 components. In order to reduce the computational demands of the problem, many simulators group the n-paraffins into a small number of pseudo-components. As a result these simulations can give physically unrealistic predictions. Figure 3 shows the danger of grouping the n-paraffins into pseudo-components. The oil and pipeline data are taken from [3].

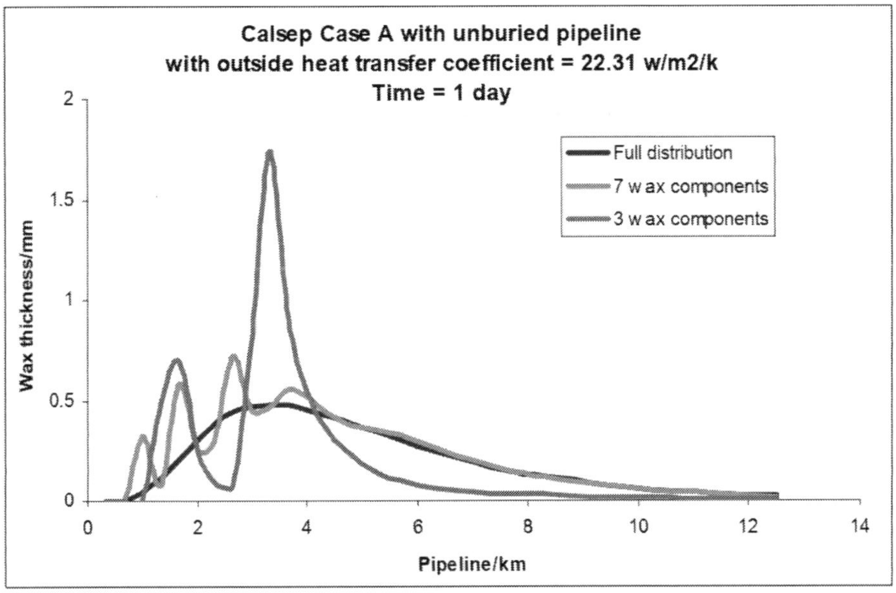


Figure 3. Wax deposition profile for simulations with different numbers of n-paraffin components

## Mass transfer rates

Fick's Law has been the most popular method of calculating wax deposition rates used by the oil industry. However, we prefer to work with mass transfer coefficients that are estimated using standard chemical engineering methods such as the Dittus-Boelter equation. Figure 4 compares predictions using these two methods with data obtained with a flow loop at the University of Tulsa. In this case the mass transfer method is in better agreement with experiment.

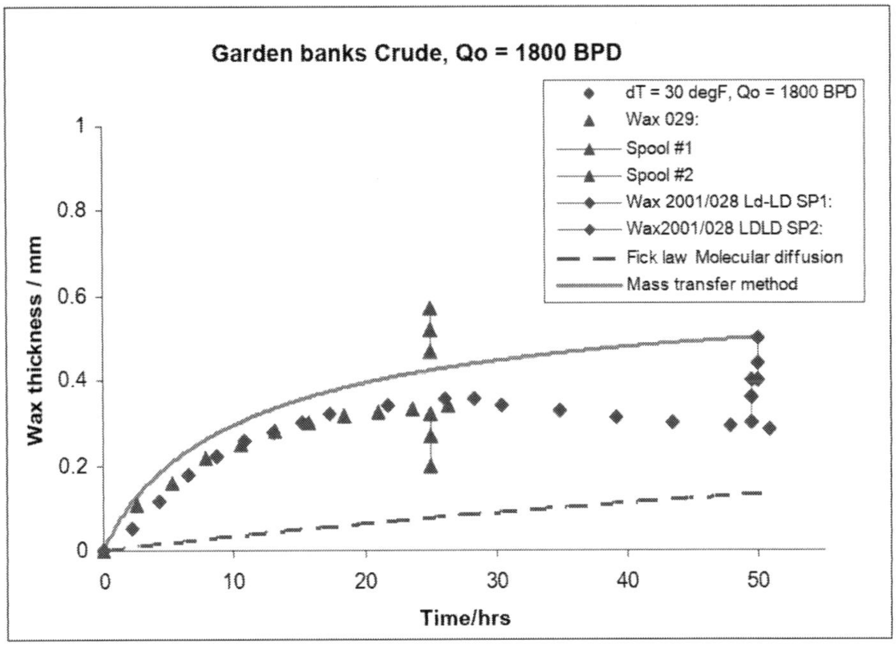


Figure 4. Comparison of mass transfer models with flowloop data

**Wax shearing**

Simple deposition calculations suggest that wax deposits tend to continue to build up in pipelines requiring remediation by, for example, frequent pigging. However, in a number of cases pipelines that operate below the WAT show no apparent wax problem. We believe this may be because the turbulence of the fluid flow tends to shear wax deposits preventing them from growing beyond a certain thickness. Figures 5a and 5b show calculations of the development of wax deposits in a real pipeline transporting a waxy condensate that, in reality, required no pigging.

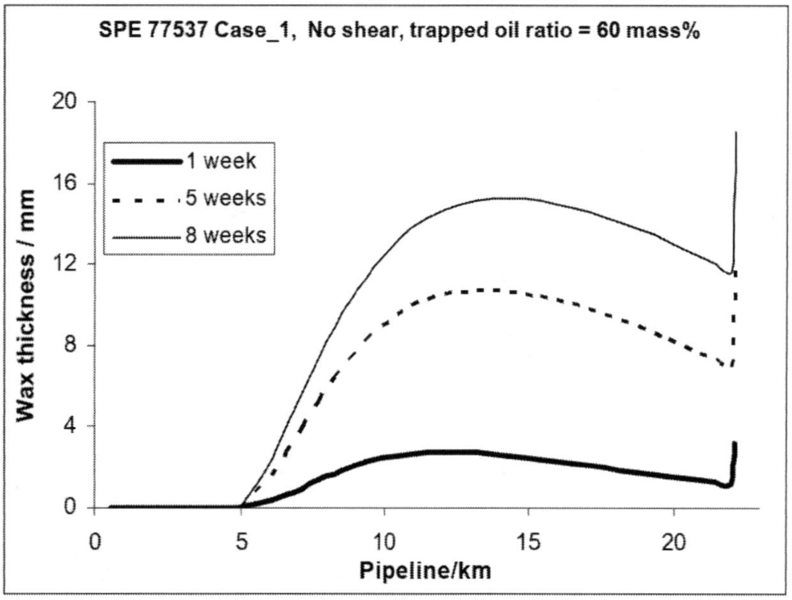


Figure 5a. Simulation of wax profile for condensate pipeline – no shearing term

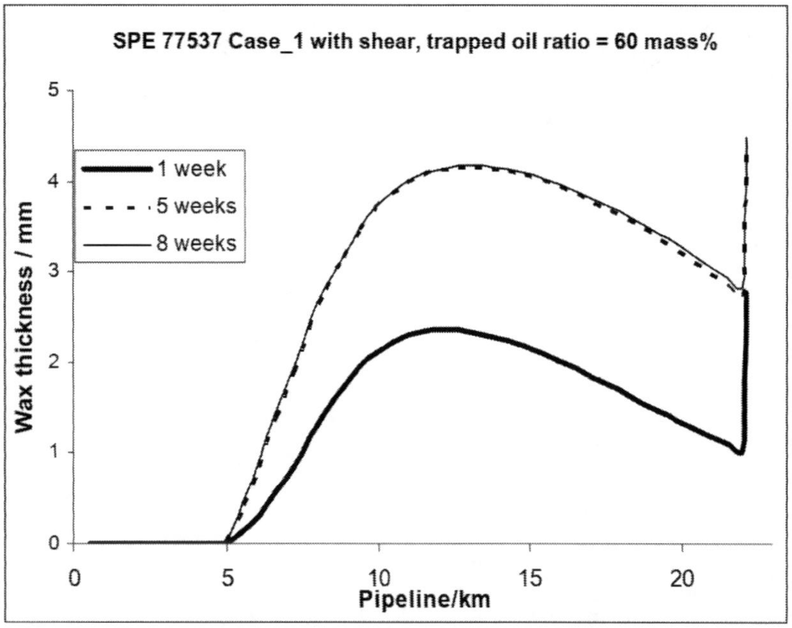


Figure 5b. Simulation of wax profile for condensate pipeline –shearing included

If wax shearing is ignored, the calculations suggest that the wax deposits grow rapidly, but if wax shearing is taken into account, the deposits cease to grow after about a month.

A detailed discussion of wax deposition modelling has been given by Moorwood et al. [4].

**Computational speed.**

To obtain physically convincing results it is important not to lump the n-paraffins of the wax into pseudo-components, but to represent all the components of the wax as a discrete distribution. To simulate wax deposition, the phase equilibria of both the flowing fluids and the deposited gel must be repeatedly calculated along with the physical properties of the phases present. The simulations are computationally intensive taking several hours for a typical real pipeline. To make the model into a practical engineering tool, the calculations have to be performed much more quickly. To do this the phase equilibrium and physical property data are represented by empirical (and compositional) expressions that are fitted to the rigorous model. Figure 6 shows a comparison between the rigorous and approximate forms of the model.

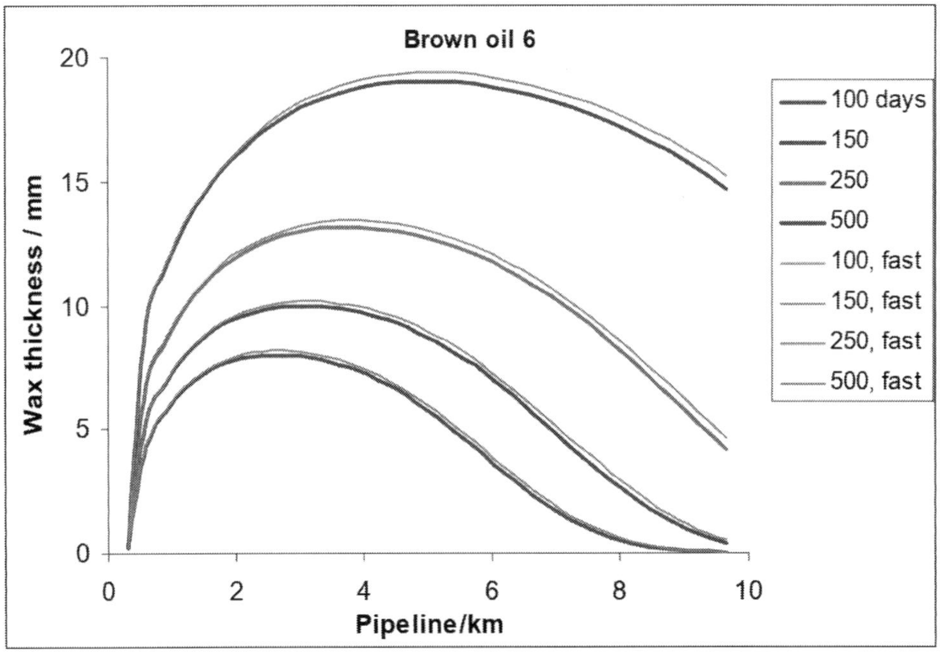


Figure 6. Comparison of simulations with rigorous and regressed (fast) property calculations

There is no significant difference in the results but the impact on run time is enormous. The rigorous version of the example (which contains 74 components) took several hours to run, whereas, using the same computer, the approximate version took about 20 seconds for the physical property correlations to be regressed, followed by the actual simulation of the pipeline which took about 1 minute.

**CAPE-OPEN interface**

The Multiflash physical property package can model multiphase equilibria involving fluid and solid phases including waxes. It has a CAPE-OPEN Thermodynamics and Physical Properties interface that conforms with the version 1.1 specification. This latest version of the specification makes it possible to describe complex phase behaviour involving many phases and including solids like wax. We have interfaced Multiflash with the FloWax simulator both directly and via a CAPE-OPEN interface. If the rigorous model is used, the impact of the CAPE-OPEN interface is to increase the run time which is already substantial. However, the approximate form is highly compatible with CAPE-OPEN as the number of physical property calls is limited and the regression stage of the simulation only lasts for a number of seconds. The actual pipeline simulation requires no direct physical property calls and therefore its run time is the same whether or not the CAPE-OPEN interface is used.

Some desirable functionality is missing from the CAPE-OPEN thermodynamics interface. When creating the property approximations they must be continuous across phase boundaries and therefore it is sometimes necessary to evaluate the properties of phases that in reality do not exist at the specified conditions. A useful tool for doing this is a phase stability test that provides a composition for incipient (metastable) phases. Although the stability test is available in Multiflash there is no means of using it via CAPE-OPEN interfaces.

A further advantage of the CAPE-OPEN interface is that it gives more flexibility when carrying out flow assurance studies. The choice of physical property model can be made independently of the deposition simulator. The FloWax program could be used with an alternative source of wax and fluid physical properties, and likewise other wax deposition programs can use the Multiflash package provided CAPE-OPEN 1.1 interfaces are implemented.

We conclude that the CAPE-OPEN 1.1 thermodynamics specification is suitable for this application but computational speed is an important consideration and interfaces must be designed with care in order to avoid unnecessary calculations and to maximise the use of available information.

**References**


1. Erickson, D.D., Niesen, V.G., Brown, T.S., *SPE 26604* (1993).
2. Coutinho, J. A. P.; Edmonds, B.; Moorwood, T.; Szczepanski, R.; Zhang, X. 'Reliable Wax Predictions for Flow Assurance', *Energy & Fuels*, *20*, 1081 (2006).
3. Lindeloff, N., Krejbjerg, K., *Energy & Fuels*, 16, 887 (2002).
4. Moorwood, T.; Zhang, X.; Edmonds, B.; Szczepanski, R. 'Simulating Wax Deposition in Pipelines for Flow Assurance', *Energy & Fuels* in press (2007).

# 459e Simulis Thermodynamics: A Cape-Open Compliant Framework for Users and Developers

**Olivier Baudouin**, Alain Vacher, and Stéphane Dechelotte. ProSim, Stratege Batiment A, BP 27210, Labege Cedex, 31672, France

Simulis® Thermodynamics is a thermophysical calculation server that generates highly accurate pure component and mixture properties (thermodynamic, transport, compressibility…) and fluid phase equilibrium (liquid-vapor, liquid-liquid and liquid-liquid-vapor). The component-oriented approach of its architecture is based on the Microsoft®'s COM/DCOM middleware. One main benefit of Simulis® Thermodynamics is its CAPE-OPEN compliance through its CAPE-OPEN thermodynamic plug and socket facilities. The CAPE-OPEN Thermodynamic specifications 1.0 and 1.1 have been both implemented and validated. Another main benefit is the capability to use legacy codes either as a DLL (Dynamic Link Library) following a standard syntax, either as VB Script (Visual Basic Script) directly written from the Simulis® Thermodynamics' environment. The standard version of Simulis® Thermodynamics is provided as an add-in in Microsoft® Excel and as a toolbox in MATLAB® and enables the user to run complete thermodynamic calculations in these applications, but it can also be plugged in any legacy code using the SDK (Software Development Kit). Also, beside the description of all these different features, several examples will be presented.

# CAPE-OPEN LOGGING AND TESTING TOOL SUPPORTS CAPE-OPEN THERMODYNAMIC INTERFACES VERSION 1.1

*Dr. Malcolm Woodman, BP Exploration Operating Company Ltd, Sunbury-on-Thames, UK*
*Peter Banks, c/o BP Exploration Operating Company Ltd, Sunbury-on-Thames, UK*
*Dr Michel Pons, CAPE-OPEN Laboratories Network, Lyon, France*

## Introduction

In 1994 BP suggested the development of a plug and play standard in process simulation software during discussions at a FOCAPD meeting held in Snowmass, Colorado. Subsequently, within the European Union (EU) funded PRIMA project (PRocess Industries Manufacturing Advantage), BP found support, particularly from ELF, BASF AG and Bayer AG, for submitting to the EU a proposal for a project aiming at process simulation software interoperability. This eventually resulted in a series of EU-funded projects, led by IFP, starting with CAPE-OPEN (Jan 1, 1997 till June 30, 1999) and followed by Global CAPE-OPEN (July 1, 1999 till March 31, 2002) and GCO Support projects. Global CAPE-OPEN was also labeled as an Intelligent Manufacturing Systems project facilitating the involvement of partners from Canada, US and Japan.

The CAPE-OPEN and Global CAPE-OPEN projects developed a set of interface standards that enabled interoperability between the various pieces of software making up process simulation tools. The GCO Support project delivered a series of disseminating means for CAPE-OPEN standards.

More than ten years after launching the idea, software developers have implemented CAPE-OPEN interfaces in many codes (1) and end-users are taking advantage of the interoperability provided by CAPE-OPEN (2). Following the end of the EU projects, the CAPE-OPEN standards are now maintained and developed by the CAPE-OPEN Laboratories Network (CO-LaN), which is funded by process industries end user companies (3).

Special Interest Groups are entities within CO-LaN that group experts of a dedicated domain and that work together according to a charter defining their scope of work and deliverables. Over a number of years the CO-LaN Special Interest Group (SIG) on Thermodynamics has reworked the interface specification for Thermodynamic and Physical Properties and the new version, numbered 1.1, has been endorsed by the CO-LaN Management Board in October 2006.

Along with its mission to maintain and further extend the CAPE-OPEN standards, CO-LaN carries the development of tools facilitating the implementation and use of CAPE-OPEN technology.

## Problem statement

The CAPE-OPEN set of interface specifications consists of a large collection of documents (4) that needs to be read, understood and implemented precisely in order to achieve interoperability. By construction CAPE-OPEN is not a piece of software: it is a documentation set freely available to the CAPE community. Some examples of implementation are available from CO-LaN but CAPE-OPEN interfaces are only specified: a developer has to code them in his/her own way. So there is no ready-to-use interface implementation supplied by CO-LaN.

As with any software development, the implementation process can lead to errors that might create interoperability mismatches between software components.

Users of process simulation tools want reliable and seamless interoperability through CAPE-OPEN technology. For end-users, using CAPE-OPEN compliant tools should not be more difficult than using components from within the generic libraries attached to any process simulator. No understanding of CAPE-OPEN technology is required to manipulate CAPE-OPEN compliant tools. However if interoperability fails, end-users need to be able to document adequately and easily what goes wrong, especially since they are often the only ones having access to all the software components involved.

Developers want to reduce the learning curve, meaning they aim at reducing the cost of adopting CAPE-OPEN. CO-LaN wants to accelerate adoption of CAPE-OPEN: getting more components and environments available with this technology implemented.

Training is an adequate solution to ease up understanding and to reduce difficulties in implementing the CAPE-OPEN technology. CO-LaN supports the training offers on CAPE-OPEN. For example a short course on CAPE-OPEN main concepts and implementation has been delivered by ProSim SA on November 12, 2006 as a Short Course within the AIChE 2006 Annual Meeting. Another short course was delivered by CO-LaN on March 7, 2007 within the $4^{th}$ CAPE-OPEN European Conference. It dealt specifically with the use of the .net framework for developing CAPE-OPEN compliant components.

However there is a need for support tools to be always available to both developers and users.

## CAPE-OPEN Logging and Testing Tool (COLTT)

The CAPE-OPEN Logging and Testing tool (COLTT) is one of the tools provided by CO-LaN to help with the issues described above.

### *COLTT architecture and design*

COLTT works with CAPE-OPEN Process Modeling Components (PMCs) and a CAPE-OPEN Process Modeling Environment (PME). Its role is to capture and record information about the interaction, through CAPE-OPEN interfaces, between a PME and a PMC (or combination of PMCs) in a form that makes it easy to detect problems or potential problems and to document them. The tool provides the user with the ability to control which combination of components is logged, how much information is logged and where the information is logged.

COLTT works only with Microsoft COM implementations of the CAPE-OPEN standards. This restriction (CORBA implementations of the CAPE-OPEN standards exist) has been applied since Microsoft COM implementations form the vast majority of all implementations these days.

COLTT intercepts the creation of objects by modifying the Windows registry so that the appropriate logger (Unit Operation, Material Object, Property Package, etc…) gets constructed instead of the requested object. The logger then constructs the requested object and forwards all calls to it. As a consequence CAPE-OPEN PMCs are not loaded differently when they are logged. The Windows registry is restored to its initial state when logging is disabled.

The logger intercepts method calls by implementing all the interfaces that a given PMC is expected to support.

COLTT generates two types of information:

- a trace of the sequence of calls made between two components, showing arguments, results and error codes; and,
- the results of checks executed by the tool to validate that calls are in the right sequence, that the

arguments are correctly implemented and that the results of the call are correctly implemented.

Configuring COLTT so that a PMC is being logged does not change the behaviour of the PMC or the PME: using a logged PMC is no different for the end-user from using a non-logged PMC. In particular:

- The lifetime of a logged component is the same whether logged or not.
- COLTT determines which of the Microsoft persistence interfaces a PMC supports and behaves accordingly when a PME requests one of the interfaces: if the PMC does not support the requested interface then at runtime COLTT does not support it either; if the PMC does support the requested interface then at runtime COLTT also supports it and logs calls to it.
- The rules of COM identity are preserved when a PMC is logged: two requests for the same interface from the same PMC returns the same pointer.

The set of CAPE-OPEN interfaces available to a PME from a PMC is not changed when the PMC is being logged. So for example, if a PMC does not support CAPE-OPEN error interfaces then a PME will not be able to access CAPE-OPEN error interfaces when the PMC is being logged.

***Development***

A COLTT prototype was developed from the end of 2004 till the beginning of 2005 to demonstrate that the concept chosen for logging was adequate. This development was undertaken by Michael Halloran, then from AspenTech, with support from SHMA Ltd and the prototype was demonstrated at the 2nd CAPE-OPEN European Conference organized by CO-LaN on February 17, 2005 in Como, Italy.

Then a specification document for a complete tool was created by AspenTech under the auspices of the Interoperability Support Special Interest Group of CO-LaN.

CO-LaN launched subsequently a development action split into a number of phases. In Phase I, forty-four combinations of PMCs and PMEs were tried out with COLTT, pinpointing a number of problems. In Phase II, the code was modified to overcome the problems found in Phase I. Phase III was begun in November 2006 to implement the complete specification defined. These three phases were contracted by CO-LaN to SHMA Ltd in Pakistan. The first production version of COLTT was released on March 8, 2007 during the 4th CAPE-OPEN European Conference.

This first version of COLTT released supported logging of thermodynamic and unit operation PMCs compliant with version 1.0 of the CAPE-OPEN standards.

***Specific development for version 1.1 of Thermodynamic and Physical Properties specification***

Version 1.1 of the CAPE-OPEN Thermodynamic and Physical Properties interface specification has been released by the CAPE-OPEN Laboratories Network (CO-LaN) in October 2006. Commercial implementations of this new version have already been released and more are planned (5). CO-LaN has decided to extend the capabilities of COLTT in order to accompany the move to version 1.1 that a number of software vendors have decided to make. This has resulted in a new version of COLTT distributed to the CO-LaN membership.

The modular architecture given to COLTT facilitates the extension to version 1.1 of the Thermodynamic and Physical Properties interface specification.

The new thermodynamic interface specification defines several new primary PMCs: Property Package Manager (PPM), Property Package (PP), Physical Property Calculator (PPC) and Equilibrium Calculator (EC). Each of these primary PMCs needs to be associated with a new logger that can be substituted for it during logging. Enabling support for version 1.1 necessitates also developing a logger for a Material Object exhibiting the specific interfaces of version 1.1. So supporting version 1.1 in COLTT means mostly adding new loggers to the code.

The Logging Controller has been modified to provide a list of PPMs and PPs compliant with version 1.1. Since a PP supports the CAPE-OPEN interfaces supported by a Physical Property Calculator or an Equilibrium Calculator, no specific list of PPCs and ECs was created at this stage. PPMs and PPs may be selected for logging with the Logging Controller.

The PP logger is able to catch communication through common CAPE-OPEN interfaces such as ICapeUtilities, ICapeIdentification as well as through the interfaces specific to version 1.1: ICapeThermoCompounds, ICapeThermoPropertyRoutine, ICapeThermoMaterialContext, ICapeThermPhases, ICapeThermoUniversalConstant, ICapeThermoEquilibriumRoutine. Figure 1 provides the interface diagram that lists all the methods logged interface per interface and specific to version 1.1.

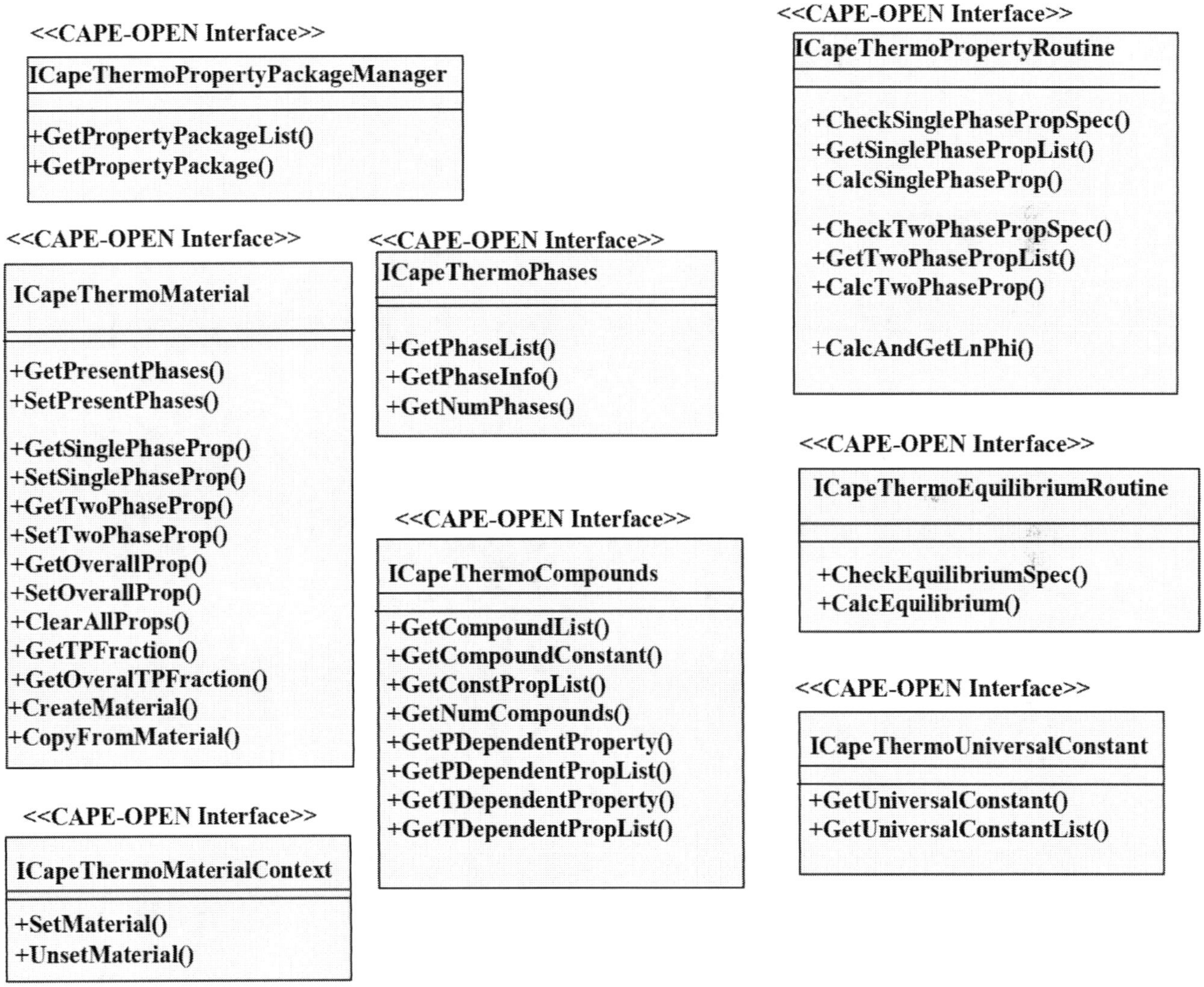


Figure 1: Thermodynamic and Physical Property interface diagram for version 1.1

Version 1.04 of COLTT contains this development. It has been tested with a number of PMEs and PMCs compliant with version 1.1 of the Thermodynamic and Physical Properties interface specification: COFE 1.07 (PME) and TEA 1.07 (PMC) from AmsterCHEM, MultiFlash 3.6 (PMC) from Infochem, an executable (PME) using a ThermoLibrary developed by ProSim SA, SolidSim

(PME) and SolidSim Property Package (PMC) from Technische Universität Hamburg-Harburg.

***Example of use***

Enabling COLTT upon a Process Modelling Component is done through the use of the CAPE-OPEN Controller and its dedicated Graphical User Interface. COLTT allows the user to configure logging for PMCs installed on the local machine only. It is possible to enable and disable logging for a specific PMC. It is also possible to find out which PMCs are being logged. Figure 2 shows that COLTT Controller is able to list all Property Package Managers and standalone Property Packages that are compliant with version 1.1 of the Thermodynamic and Physical Properties interface specification.

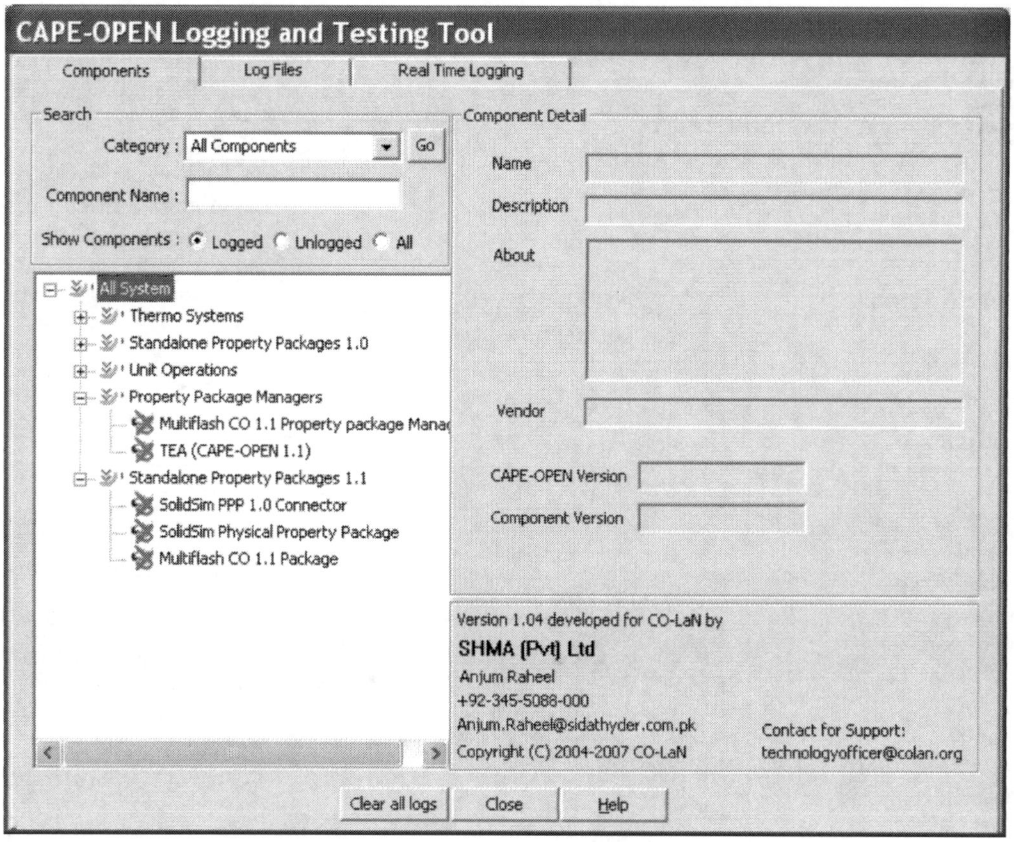


Figure 2: COLTT controller

The user selects which PMC(s) need to be logged. In Figure 3 Multiflash CO 1.1 Property Package Manager is selected for logging. The PMC components that COLTT presents for logging are the primary CAPE-OPEN components that a user can select within a PME. Secondary CAPE-OPEN components such as errors, ports, parameters and Material Objects are logged automatically as a consequence of logging a primary PMC. Here all Property Packages under the MultiFlash Property Package Manager (PPM) will have their communication logged when they are next used, whatever the PME that uses them.

A simulation run is then conducted in COFE, AmsterCHEM's process simulator that offers compliance with CAPE-OPEN thermodynamic and unit operation components. A simple flash calculation is performed as described in Figure 4.

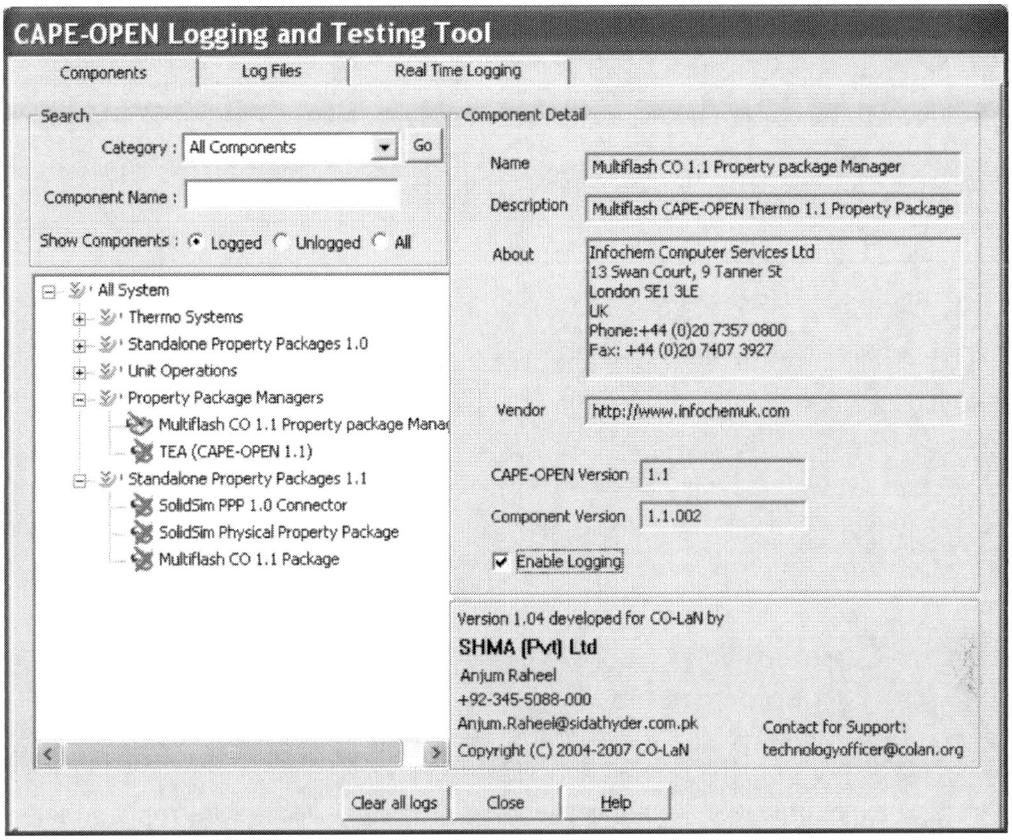


Figure 3: Enabling logging on MultiFlash Property Package Manager

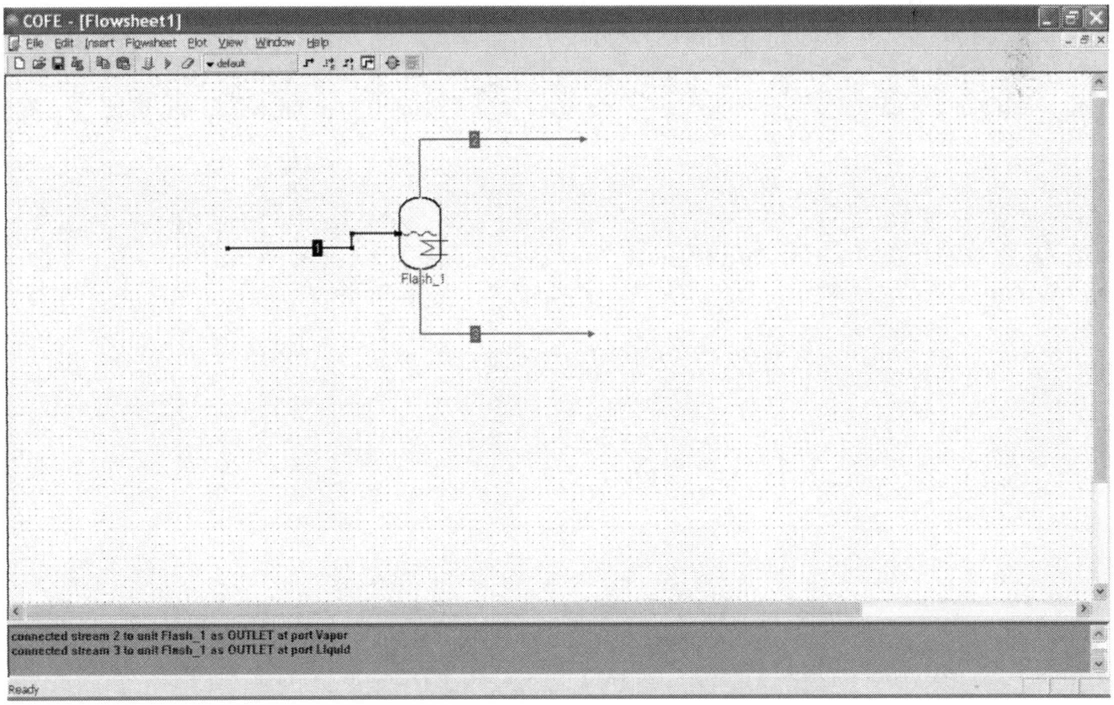


Figure 4: screenshot of run in COFE 1.08 with MultiFlash Property Package

***Analysis of log file***

At the end of the run, the log file created by COLTT can be displayed. By default, COLTT logs all calls made in both directions, via CAPE-OPEN interfaces between a PME and a PMC.

```
PropertyPackage <C3C5CO> : Call to GetTDependentPropList
  GetTDependentPropList Returns:
    vaporPressure
    vaporPressure.Dtemperature
    idealGasEnthalpy
    idealGasEntropy
    idealGasHeatCapacity
    idealGasHeatCapacity
    volumeOfLiquid
    volumeOfLiquid.Dtemperature
    heatOfVaporization
    heatOfVaporization.Dtemperature
    viscosityOfLiquid
    viscosityOfVapor
    thermalConductivityOfLiquid
    thermalConductivityOfVapor
    surfaceTensionSatLiquid
    heatCapacityOfLiquid
    heatCapacityOfSolid
    virialCoefficient
PropertyPackage <C3C5CO> : Return from GetTDependentPropList - No Error
```

Figure 5: Example of communication logged on GetTDependentPropList method

Each call to any method from any of the interfaces generates a log entry showing:

- Which object made the call
- Which call was made
- The values for the input arguments that were passed
- The return values that were passed back

Whether the call generated an error and what the error was – error codes are explained by a message where possible, or at least translated to a Windows or CAPE-OPEN error name such as E_FAIL or ECapeLimitedImpl.

Log files use a human-readable text format so that they can be viewed easily. Figure 5 shows the logging of the call made by the client (here the Unit Operation) on the Property Package. The Unit Operation requests the list of all temperature dependent properties supported by the Property package. Figure 6 demonstrates that both input arguments as well as output arguments are logged.

```
PropertyPackage <C3C5CO> : Call to GetPhaseInfo
  GetPhaseInfo takes input Arguments:
    Phase    PhaseAttribute
    LIQUID1  StateOfAggregation
  GetPhaseInfo Returns:
    Phase    PhaseAttribute     ValueReturned
    LIQUID1  StateOfAggregation Liquid
PropertyPackage <C3C5CO> : Return from GetPhaseInfo - No Error
```

Figure 6: Example of communication logged on GetPhaseInfo method

## Conclusion and perspectives

COLTT has been upgraded to log PMCs compliant with version 1.1 of the Thermodynamic and Physical Properties CAPE-OPEN interface specification. It has been tested against a small number of PMCs and PMEs supplied by 3rd parties. It can be put to use in the near future by organizations developing PMCs and PMEs supporting version 1.1, namely TUV NEL Ltd for PPDS, Aspentech for Aspen Hysys and COMThermo, The Tulsa University for TUWAX.

COLTT is continuously maintained and enhanced by CO-LaN. Among the future enhancements planned, a change of output format is being considered, using XML to provide an easier parsing through the information logged by COLTT.

## References


1. Pons, Michel (2003), AIDIC Conference Series, vol. 6: Selected papers of the Sixth Italian Conference on Chemical and Process engineering, pp. 253-262. S. Pierucci (ed.), Milan, Reed Business Information S.p.A., Milan.
2. Holmes J.W., D. L. Johnson, S. Ogundiran and M. Lichtenauer (2006), 06'AIChE, Paper 535c, in CDRom, San Francisco, CA.
3. Belaud J.P. et al. (2003), Proceedings of Foundations Of Computer-Aided Process Operations, Coral Springs, Florida, USA.
4. Belaud, J.P. and M. Pons (2002), Computer-Aided Chemical Engineering, vol. 10: Proceedings of the 12th European Symposium on Computer Aided Process Engineering, pp 847-852, J. Grievink and J. van Schijndel (eds.), Elsevier Science B.V., Amsterdam.
5. Pons, Michel (2007), The Eight International Conference on Chemical and Process Engineering, S. Pierucci (Ed.), AIDIC, Milan.

# 459g Parameter Estimation in Gproms Using Aspen Properties

**Kevin C. Seavey** and Wilma Hensen. Engineering Solutions, The Dow Chemical Company, 2301 Brazosport Blvd, APB Building, Freeport, TX 77541-3257

In this work, we use gPROMS (PSE, London, UK) to characterize a large number of Arrhenius constants using many sets of liquid-vapor, semi-batch reaction data. The components and phase equilibrium were previously characterized using the electrolyte non-random, two-liquid property method in Aspen Plus (Aspen Technology, Cambridge, MA, USA). Our goal was to determine the reaction constants in a manner consistent with the predefined physical properties and vapor-liquid equilibrium. We encountered several difficulties when developing a gPROMS model that made calls to Aspen properties through the CAPE-OPEN interface. The principal difficulty was that simulating one reaction experiment took 11 minutes of CPU time. For us, this meant that a single parameter estimation run took about nine days. We found that more than 98% of the CPU time was taken up by the liquid-phase property calculations through the CAPE-OPEN interface, such as those for the density and the fugacity coefficient. To work around this problem, we developed empirical expressions fitted to data generated by the property method itself. We implemented these expressions directly in the model code, freeing it from having to make calls to Aspen Properties via the CAPE-OPEN interface. By doing so, we were able to reduce the model run time by more than a factor of 50.

## 524a Interoperability Workshop Contents

**Michel Pons**, CAPE-OPEN Laboratories Network, 32, rue Raulin, F-69007, LYON, France

The Interoperability Workshop is intended to permit developers to test their CAPE-OPEN compliant PMCs and PMEs and to enable end-users to have a first hand grasp of what CAPE-OPEN is.

Web Page: www.colan.org

# AUTHOR INDEX

**American Institute of Chemical Engineers**
3 Park Avenue
New York, New York 10016-5991

**ISBN: 978-1-60423-840-2**